# FRACTION CIRCLE
## ACTIVITIES

Barbara Berman

Fredda J. Friederwitzer

DALE SEYMOUR PUBLICATIONS

Cover design: Rachel Gage

ISBN 0-86651-432-5
Order number DS15800

DALE
SEYMOUR
PUBLICATIONS
P.O. BOX 10888
PALO ALTO, CA 94303

10 11 12 13 14 15-MA-95

# Introduction

## Why do we need manipulatives?

The use of manipulatives in teaching mathematics is not new. The first concrete manipulative devices were probably the fingers, toes, and pebbles used by primitive people to keep a record of important information. Since then, educators like Pestalozzi, Montessori, and Dewey have repeatedly urged the use of concrete objects to develop understanding of abstract concepts. Recently, Piaget focused attention on the need for children to experiment and discover mathematical principles themselves. He suggested that a student's failure to learn may be caused by a "too-rapid passage" from concrete experiences to abstract words and symbols.

Many teachers consider fractions among the most difficult mathematical topics to teach. Yet children come to school with a wealth of experience and a variety of fraction concepts developed through the unplanned, spontaneous activities of daily life. While some of these concepts are accurate, some are completely erroneous; for example, "the bigger half of the apple."

The experiences children have in school, however, *can* be carefully planned. When these experiences are structured along a concrete-to-abstract continuum, using appropriate manipulative models to embody the concepts being taught, children can easily learn very complex ideas.

## Using manipulative fraction circles and this book

The activities in this book help students develop fraction concepts through the use of a manipulative model—fraction circles. Using fraction circles and these sequentially-developed activities, children will discover:

- names for unit fractions
- different names for "one"
- the relationship between the size of a fraction piece and the written symbol used as its denominator
- the relationship between improper fractions and mixed numbers
- equivalent fractions
- addition and subtraction of fractions with like and unlike denominators
- lowest terms

Please note that pages 2-10, which introduce the different fraction circle pieces, do not have to be done in sequence; the remaining activities are designed to be used sequentially.

Manipulative fraction circles consist of different colored pieces representing halves, thirds, quarters, sixths, eighths, and ninths. They are available from educational suppliers. Each student should have a set of fraction circle pieces to do the activities in this book. For these activities they will also need a whole (or unit) circle, fifths, tenths, and twelfths. Since these pieces are not included in the standard sets, we have provided cutouts for them on the back cover foldout. You will probably find it easiest to cut out these pieces ahead of time, or duplicate them and have your students paste them on tagboard and color them before beginning the activities.

The activities are designed to be used in a variety of settings: whole group, small group, with a partner, and individually. The cooperative small group of three or four students is particularly appropriate since this configuration encourages the processes of discussing, sharing, experimenting, and discovering, which are used throughout the book. Of course, the discretion of the classroom teacher and the needs of the students dictate the setting employed. An answer key for each activity is provided beginning on page 46.

With any structured manipulative activity, children need time to explore the material and generate their own discoveries. They also need time to investigate the problems, develop solutions, and discuss and share ideas with others. Research shows that time spent on initial concept development reduces the amount of time needed later for drill.

Students should manipulate the fraction circles as they solve each set of problems. Eventually, when they are confident about understanding the ideas, children will stop reaching for the fraction circle pieces. Experiences with the fraction circles then become the basis for an understanding of how to use fractions as numbers.

# The Unit Circle

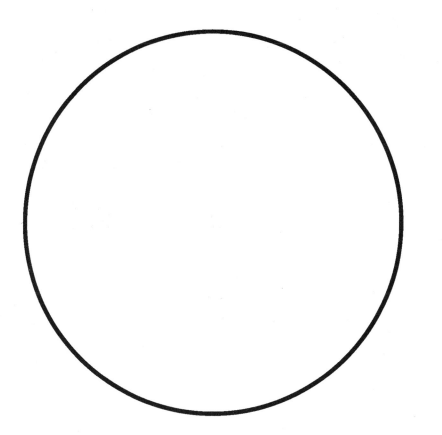

This is a *unit circle.* Another name for this unit circle is *1 whole* or *1.*

Find the piece that matches this picture. What color is it? _____

This unit circle can be cut into many different parts.
You will use the colored fraction circle pieces to learn about these parts.

# Two Equal Parts

1. Find this piece. What color

   is it? _____

   Are all the pieces of this color

   the same size? _____
   Prove it.

2. Use as many pieces of this
   color as you can to cover the
   *unit circle.*

   How many pieces did you use

   to cover the unit circle? _____

3. One of these pieces is 1 out of
   2 needed to cover the unit
   circle.

   We can call it "1 out of 2" or

   $\frac{"1"}{2}$ or "one-half" of the whole
   circle.

   Another name for the unit
   circle could be _____ halves.

   **We can say 1 unit = 2 out of 2 = $\frac{2}{2}$.**

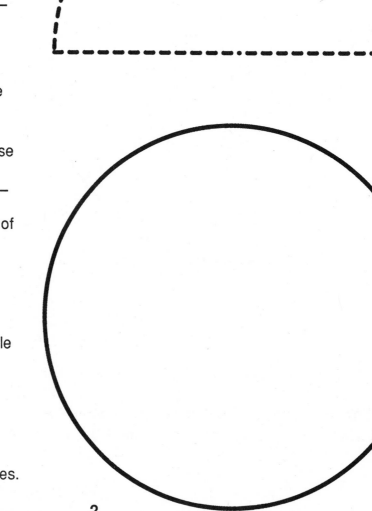

# Three Equal Parts

**1.** Find this piece.  What color

   is it? _____

   Are all the pieces of this color

   the same size? _____

   Prove it.

**2.** Use as many pieces of this
   color as you need to cover the
   unit circle.

   How many pieces did you use

   to cover the unit circle? _____

   1 whole = _____ pieces

**3.** One of these pieces is

   1 *out of* _____   or   $\dfrac{\square}{\square}$   or

   *one-third* of the whole circle.

**4.** Think of a name …

   For 2 pieces: _____

   For 3 pieces: _____

**5.** 1 unit = _____ out of _____ = $\dfrac{\square}{\square}$

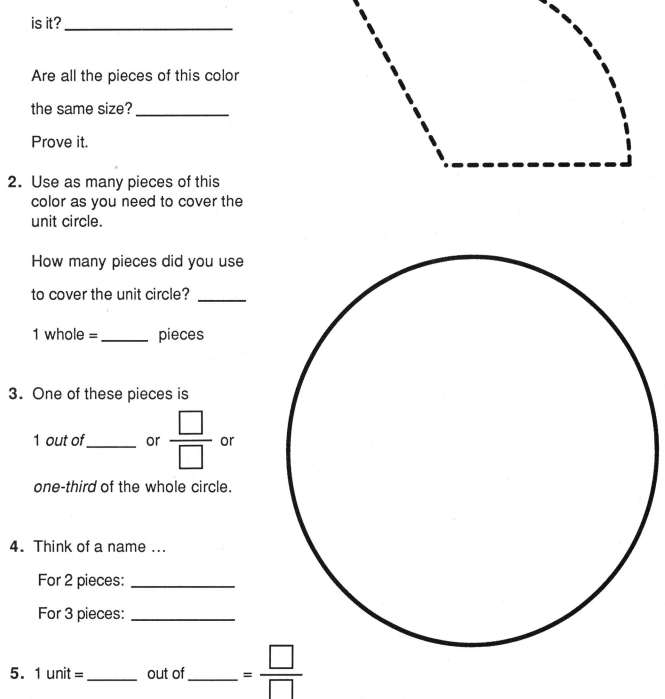

# Four Equal Parts

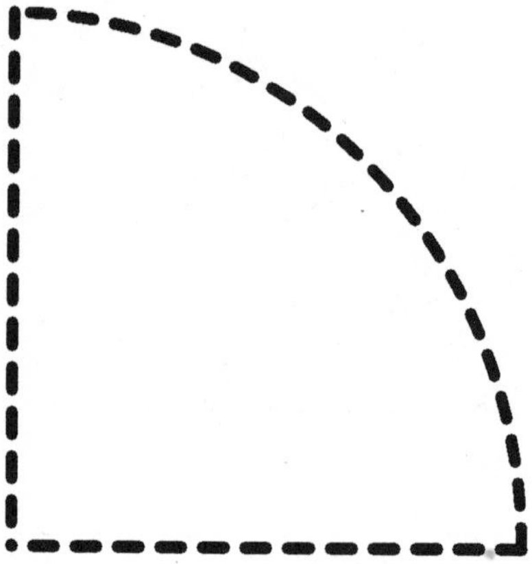

1. Find this piece. What color

   is it? _____

   Are all the pieces of this color

   the same size? _____

   Prove it.

2. Use as many pieces of this
   color as you need to cover the
   unit circle.

   How many pieces did you use

   to cover the unit circle? _____

   1 whole = _____ pieces

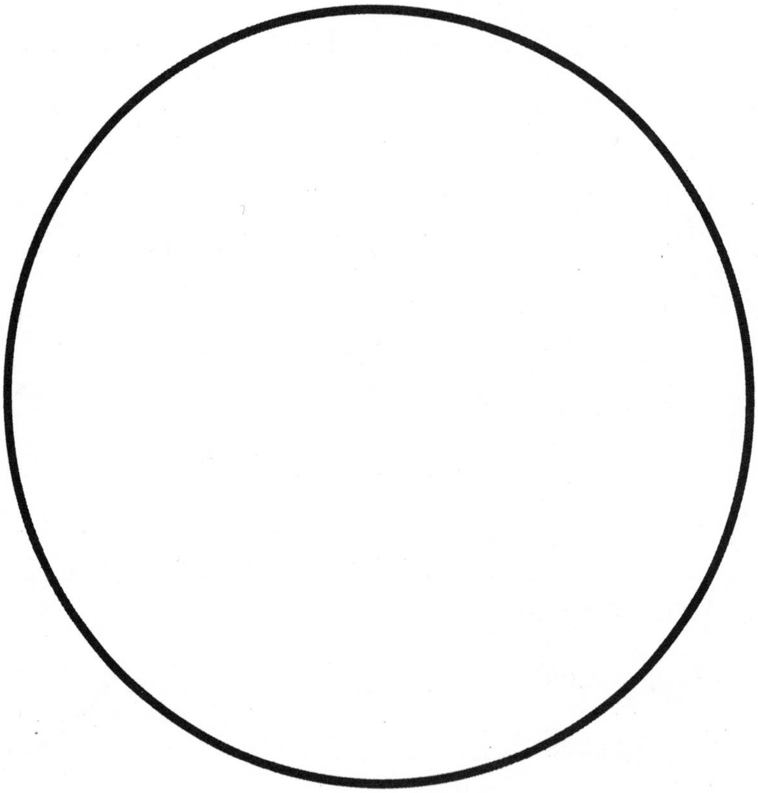

3. One of these pieces is

   1 *out of* _____ or $\dfrac{\Box}{\Box}$ or

   *one-fourth* of the whole circle.

4. Think of a name ...

   For 2 pieces: _____

   For 3 pieces: _____

   For 4 pieces: _____

5. 1 unit = _____ out of _____ = $\dfrac{\Box}{\Box}$

# Five Equal Parts

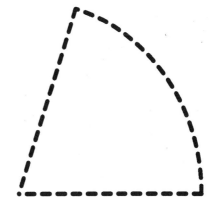

1. Find this piece.  What color

   is it? _____

   Are all the pieces of this color

   the same size? _____

   Prove it.

2. Use as many pieces of this color as you need to cover the unit circle.

   1 whole = _____ pieces

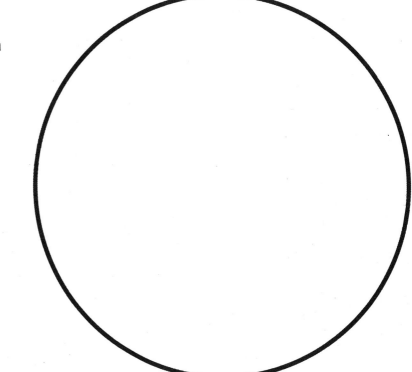

3. Using numbers, write a fraction name for one of these pieces.

   _____

   Write a word name for the

   piece. _____

   Explain why you chose

   that name. _____

   _____

   _____

4. Name 2 pieces: _____

   3 pieces: _____

   4 pieces: _____

   5 pieces: _____

5. 1 unit = _____ out of _____ = $\dfrac{\Box}{\Box}$

# Six Equal Parts

1. Find this piece.  What color

   is it? _____

   Are all the pieces of this color

   the same size? _____

   Prove it.

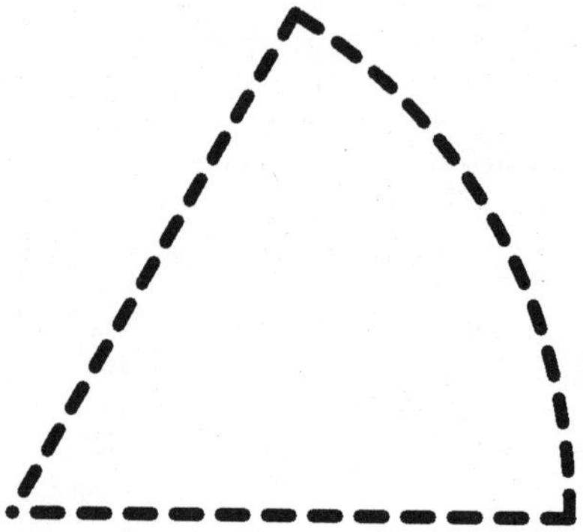

2. Use as many pieces of this
   color as you need to cover the
   unit circle.

   1 whole = _____ pieces

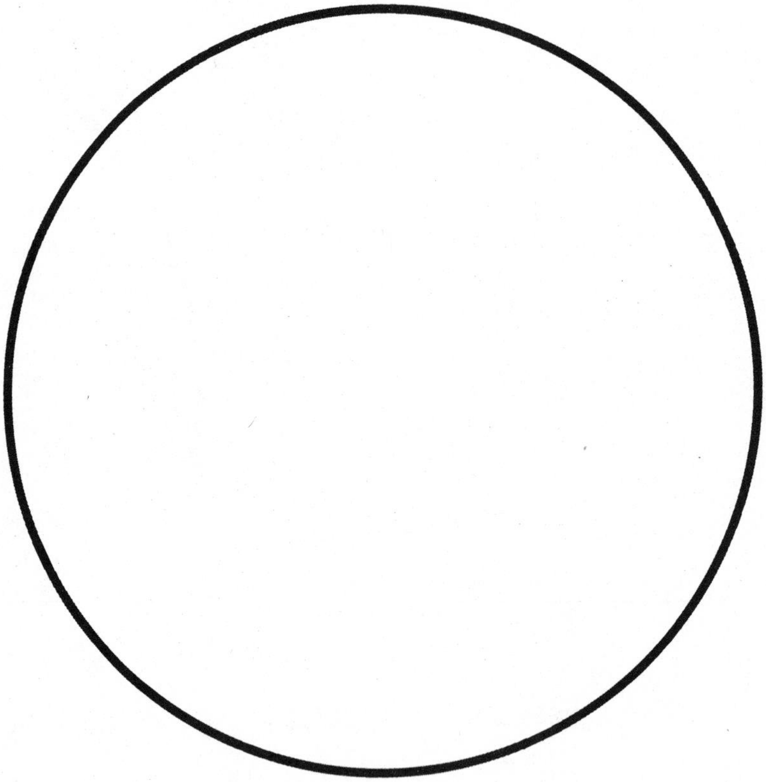

3. Write a fraction name for one of
   these pieces, using numbers
   and words:

   _____ or _____

   Explain why you chose

   that name. _____

   _____

   _____

4. Name 2 pieces: _____

         3 pieces: _____

         4 pieces: _____

         5 pieces: _____

         6 pieces: _____

5. 1 unit = _____ out of _____ = $\dfrac{\square}{\square}$

# Eight Equal Parts

1. Find this piece. What color

   is it? _____

   Are all the pieces of this color

   the same size? _____

   Prove it.

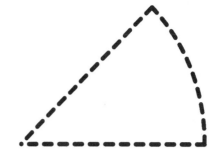

2. Use as many pieces of this color as you need to cover the unit circle.

   1 whole = _____ pieces

3. Write a fraction name for one of these pieces, using numbers and words:

   _____ or _____

   Explain why you chose

   that name. _____

   _____

   _____

4. Name 2 pieces: _____

   3 pieces: _____

   4 pieces: _____

   5 pieces: _____

   6 pieces: _____

   7 pieces: _____

   8 pieces: _____

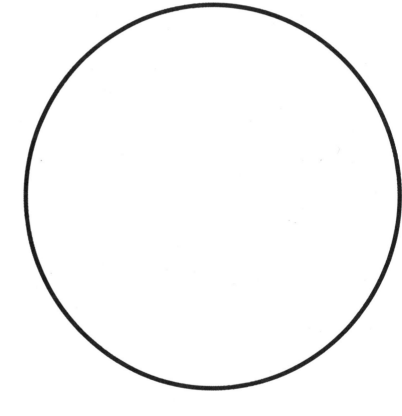

5. 1 unit = _____ out of _____ = $\dfrac{\square}{\square}$

# Nine Equal Parts

1. Find this piece. What color

   is it? _____

   Are all the pieces of this color

   the same size? _____

   Prove it.

2. Use as many pieces of this
   color as you need to cover the
   unit circle.

   1 whole = _____ pieces

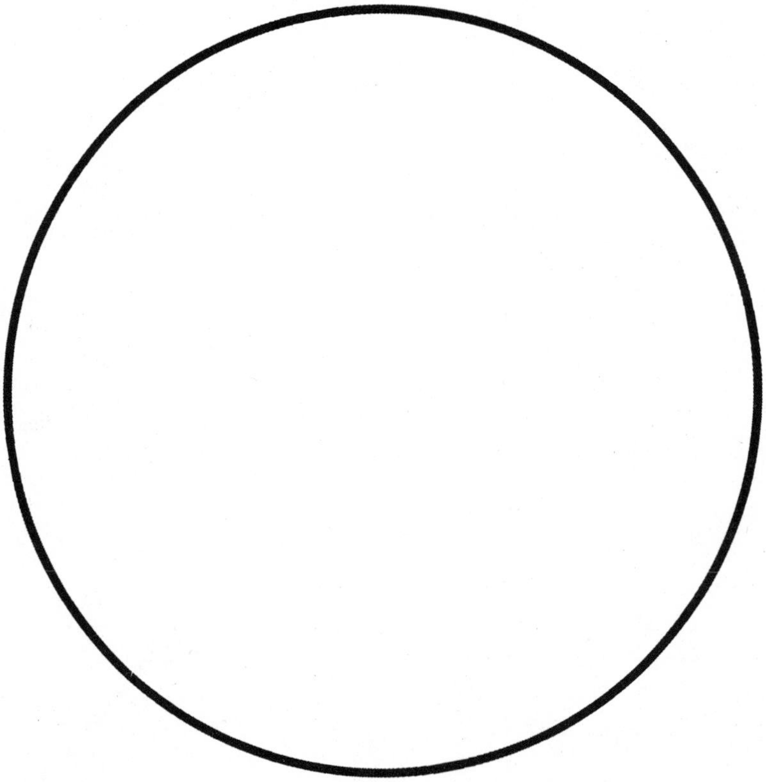

3. Write a fraction name for one of
   these pieces, using numbers
   and words:

   _____ or _____

   Explain why you chose

   that name. _____

   _____

   _____

4. Name 2 pieces: _____

         3 pieces: _____

         4 pieces: _____

         5 pieces: _____

         6 pieces: _____

         7 pieces: _____

         8 pieces: _____

         9 pieces: _____

5. 1 unit = _____ out of _____ = $\dfrac{\Box}{\Box}$

# Ten Equal Parts

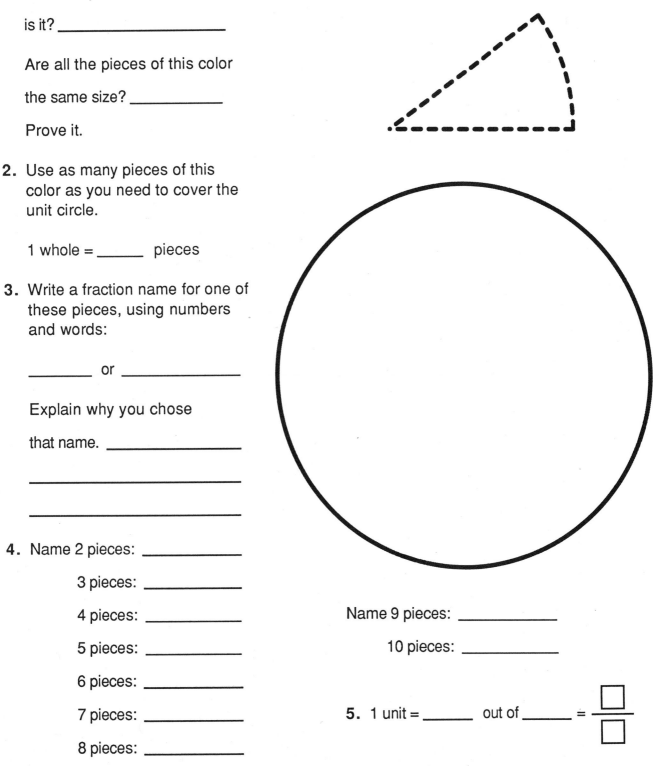

1. Find this piece. What color

   is it? _____

   Are all the pieces of this color

   the same size? _____

   Prove it.

2. Use as many pieces of this color as you need to cover the unit circle.

   1 whole = _____ pieces

3. Write a fraction name for one of these pieces, using numbers and words:

   _____ or _____

   Explain why you chose

   that name. _____

   _____

   _____

4. Name 2 pieces: _____

   3 pieces: _____

   4 pieces: _____

   5 pieces: _____

   6 pieces: _____

   7 pieces: _____

   8 pieces: _____

   Name 9 pieces: _____

   10 pieces: _____

5. 1 unit = _____ out of _____ = $\dfrac{\square}{\square}$

# Twelve Equal Parts

1. Find this piece.  What color

   is it? _____

   Are all the pieces of this color

   the same size? _____

   Prove it.

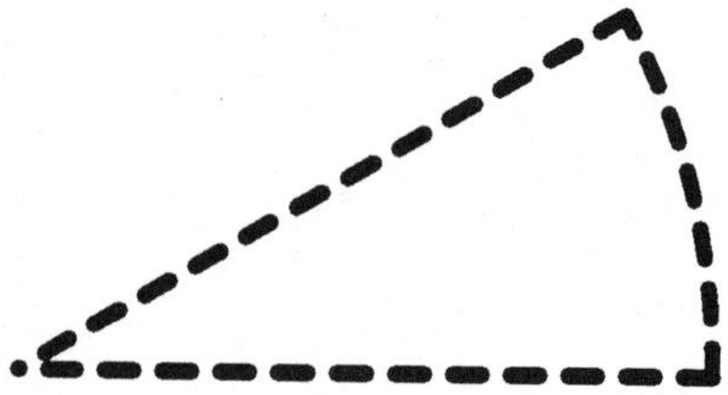

2. Use as many pieces of this color as you need to cover the unit circle.

   1 whole = _____ pieces

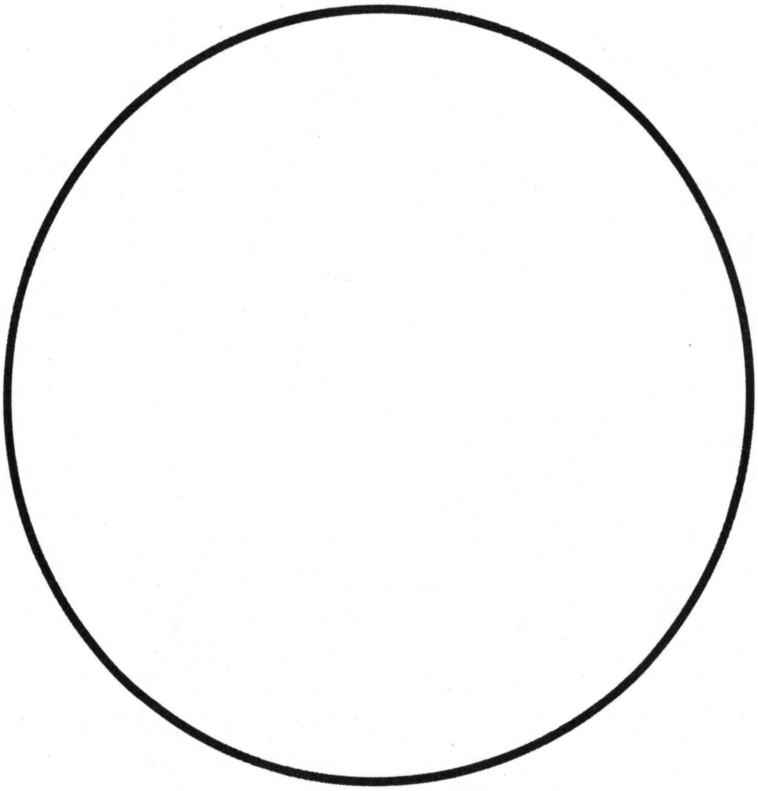

3. Write a fraction name for one of these pieces, using numbers and words:

   _____ or _____

   Explain why you chose

   that name. _____

   _____

   _____

4. Name 2 pieces: _____

   3 pieces: _____

   4 pieces: _____

   5 pieces: _____

   6 pieces: _____

   7 pieces: _____

   8 pieces: _____

   Name 9 pieces: _____

   10 pieces: _____

   11 pieces: _____

   12 pieces: _____

5. 1 unit = _____ out of _____ = $\dfrac{\square}{\square}$

# Names for 1

1. Record on the chart:
   - The number of pieces of *each color* that make 1 whole.
   - The fraction name in words and numbers for 1 piece of each color.

| Names for 1 | | | |
|---|---|---|---|
| Color | Number of pieces to make 1 | Fraction Name Numbers | Words |
|  |  |  |  |
|  |  |  |  |
|  |  |  |  |
|  |  |  |  |
|  |  |  |  |
|  |  |  |  |
|  |  |  |  |
|  |  |  |  |
|  |  |  |  |

2. There is a pattern in the fraction names for 1. Can you find it?

   Write a rule for naming 1 as a fraction. _____

   _____

   _____

# Numerator and Denominator

1. Find this piece. Think of the pieces of this color as part of a family.

   Use as many pieces as you need to cover the unit circle. How many members (or parts)

   are in this family? _____

2. When all the parts are together, they form 1 *whole* family.

   We can say that _____ out of

   ____ or $\dfrac{\Box}{\Box}$ are home.

3. Suppose that only three members of the family are home. Use numbers to show

   this as a fraction. _____

4. Which part of the fraction tells you the number of members in

   the family? _____

   This part is called the *denominator.* You can remember this word by thinking that the denominator is "down below":

   _____
   Denominator

5. Which part of the fraction tells you how many family members are

   home? _____ This part is called the *numerator.*

   $$\dfrac{\text{Numerator}}{\text{Denominator}} = \dfrac{\text{how many are home}}{\text{the number in the family}}$$

# Comings and Goings

This unit circle represents a holiday party.  Many families were invited.
Not everyone from each family went.

*For example:*  The 12-member Orange family was invited and 10 went:  $\frac{10}{12}$

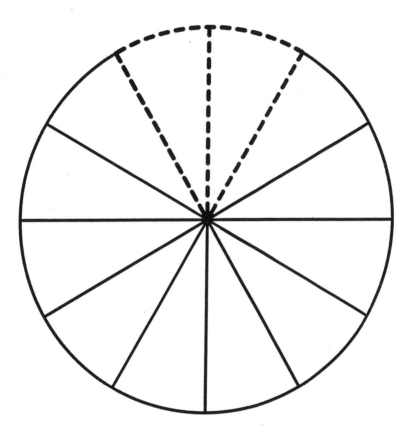

Use your fraction circle pieces to set up similar problems.  Ask a friend to write
the fraction that shows how many members of each family attended the party.

*My problems:*

# Recognizing Fractions

Find the fraction circle piece that covers each drawing below.
Write its name on the line.

# Comparing Fractions

1. Take one piece of each color.  Do *not* use the unit circle piece.
   Put them in order from *largest* to *smallest*.
   Write the name of each fraction in the answer blanks below.

_____,  _____,  _____,  _____,  _____,  _____,  _____,  _____,  _____
Largest                                                                                                    Smallest

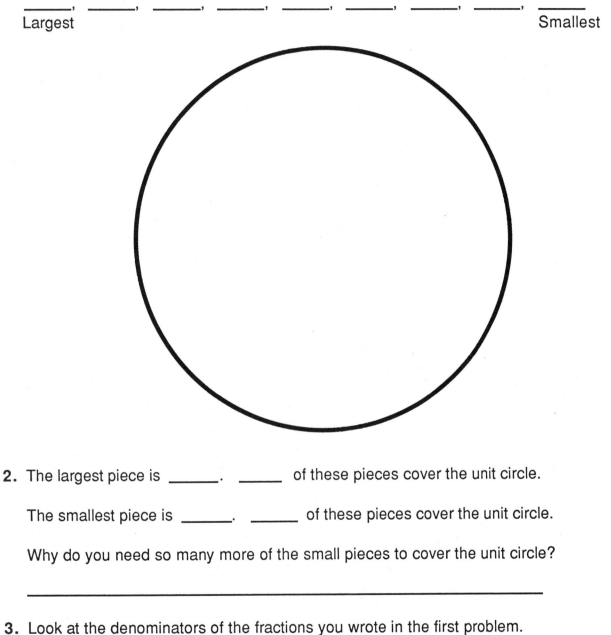

2. The largest piece is _____.  _____ of these pieces cover the unit circle.

   The smallest piece is _____.  _____ of these pieces cover the unit circle.

   Why do you need so many more of the small pieces to cover the unit circle?

   _____

3. Look at the denominators of the fractions you wrote in the first problem.
   What do you notice about the denominators as the fractions get smaller?

   _____

   Why is this happening? _____

# Which Is Larger?

Use your fraction circle pieces to compare each pair of fractions below.
Circle the *larger* fraction.

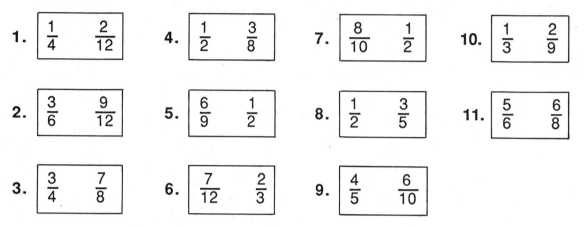

1. $\frac{1}{4}$   $\frac{2}{12}$     4. $\frac{1}{2}$   $\frac{3}{8}$     7. $\frac{8}{10}$   $\frac{1}{2}$     10. $\frac{1}{3}$   $\frac{2}{9}$

2. $\frac{3}{6}$   $\frac{9}{12}$     5. $\frac{6}{9}$   $\frac{1}{2}$     8. $\frac{1}{2}$   $\frac{3}{5}$     11. $\frac{5}{6}$   $\frac{6}{8}$

3. $\frac{3}{4}$   $\frac{7}{8}$     6. $\frac{7}{12}$   $\frac{2}{3}$     9. $\frac{4}{5}$   $\frac{6}{10}$

The symbol > is used to show *greater than* and the symbol < is used to show *less than*. An easy way to remember where to use < and > is to think of a shark. The shark always has an open mouth and swims toward the *larger* amount of food.

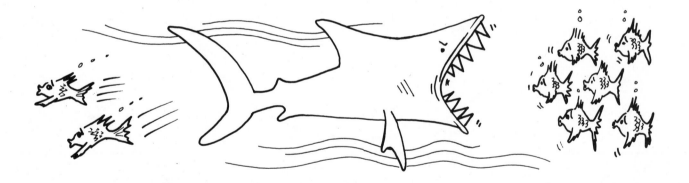

No matter which side the larger amount is on, the "open mouth" faces it. Since we read from left to right, > means *greater than* and < means *less than*.

*For example:*

3 > 1          1 < 3

3 is greater than 1      1 is less than 3

Insert the symbol > or < between the fractions in each problem at the top of this page to show which fraction is larger.

# Greater Than and Less Than

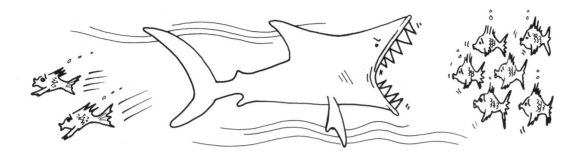

Remember, the shark always swims toward the larger amount.
Use your fraction circle pieces to compare each pair of fractions below.
Decide which fraction is larger. Insert the symbol > or < between the fractions.

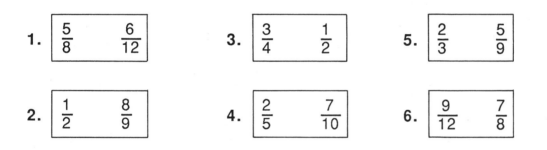

1. $\dfrac{5}{8}$    $\dfrac{6}{12}$        3. $\dfrac{3}{4}$    $\dfrac{1}{2}$        5. $\dfrac{2}{3}$    $\dfrac{5}{9}$

2. $\dfrac{1}{2}$    $\dfrac{8}{9}$        4. $\dfrac{2}{5}$    $\dfrac{7}{10}$        6. $\dfrac{9}{12}$    $\dfrac{7}{8}$

Insert the symbol >, <, or = between the fractions to show which fraction is larger, or to show they are the same size.

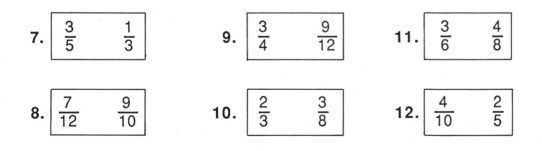

7. $\dfrac{3}{5}$    $\dfrac{1}{3}$        9. $\dfrac{3}{4}$    $\dfrac{9}{12}$        11. $\dfrac{3}{6}$    $\dfrac{4}{8}$

8. $\dfrac{7}{12}$    $\dfrac{9}{10}$        10. $\dfrac{2}{3}$    $\dfrac{3}{8}$        12. $\dfrac{4}{10}$    $\dfrac{2}{5}$

# Names for $\frac{1}{2}$

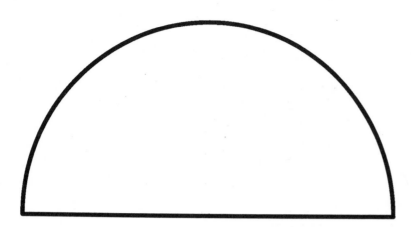

This piece = _____ of the unit circle.
Write its fraction name on the top line of the chart below.

Find five different names for this piece by covering it *exactly* with pieces of another color and counting them. Record your findings on the chart below.

| Fraction: | | | |
|---|---|---|---|
| Number of Pieces to Cover It | Color | Fraction | Different Fraction Names for ⌒ |
| 2 | brown * | $\frac{1}{4}$ | $\frac{2}{4}$ |
| | | | |
| | | | |
| | | | |
| | | | |

\* The color of this piece may be different in your fraction circle pieces set.

# Names for Other Fractions

1. This piece = _____ of the unit circle.
   Write its fraction name on the top line of the
   chart below.

   Find three different names for this piece by
   covering it *exactly* with pieces of another color
   and counting them.  Record your findings on the
   chart below.

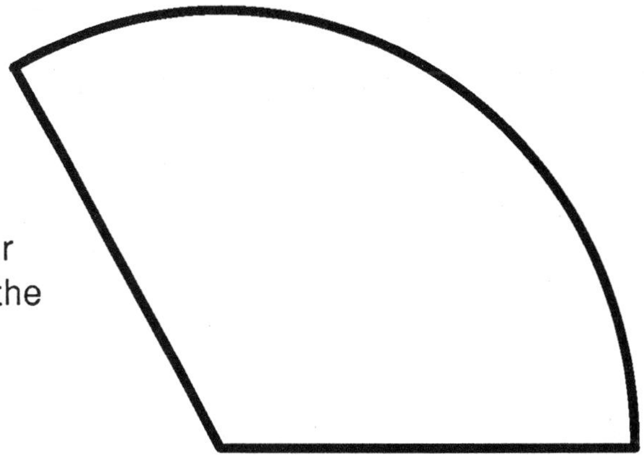

| Fraction: | | | |
|---|---|---|---|
| Number of Pieces to Cover It | Color | Fraction | Different Fraction Names for  ◁ |
| | | | |
| | | | |
| | | | |

2. What part of the unit circle is this piece? _____
   Write its fraction name on the top line of the chart below.

   Find two different names for this piece by covering
   it *exactly* with pieces of another color and counting
   them.  Record your findings on the chart below.

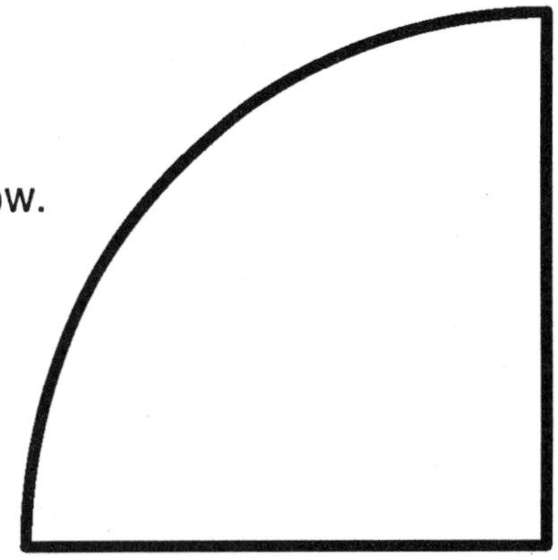

| Fraction: | | | |
|---|---|---|---|
| Number of Pieces to Cover It | Color | Fraction | Different Fraction Names for ◁ |
| | | | |
| | | | |

# What's My Name?

Use your fraction circle pieces to help you find other names for these fractions. These are called *equivalent fractions*.

1. $\frac{1}{2}$ = _____
   = _____
   = _____
   = _____
   = _____

2. $\frac{2}{3}$ = _____
   = _____
   = _____

3. $\frac{3}{4}$ = _____
   = _____
   $\frac{2}{6}$ = _____
   = _____
   = _____

4. $\frac{2}{8}$ = _____
   = _____

5. $\frac{1}{6}$ = _____

6. $\frac{5}{6}$ = _____

7. $\frac{2}{5}$ = _____

8. $\frac{6}{10}$ = _____

9. $\frac{4}{5}$ = _____

10. $\frac{2}{12}$ = _____
    $\frac{3}{12}$ = _____
    $\frac{10}{12}$ = _____

BONUS:

11. $\frac{12}{4}$ = _____
    $\frac{12}{6}$ = _____
    $\frac{10}{5}$ = _____

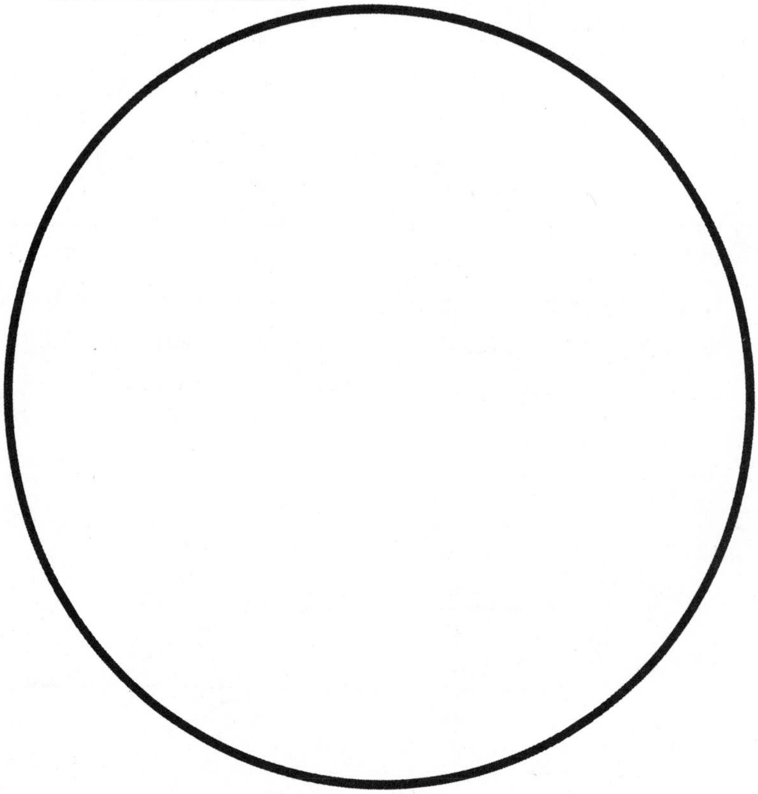

# More Equivalent Fractions

List all the names you can find for each fraction.
Use your fraction circle pieces to help you.

1.  1 = _____ = _____ = _____ = _____ = _____ = _____

    = _____ = _____ = _____

2.  $\frac{1}{2}$ = _____ = _____ = _____ = _____ = _____

3.  $\frac{1}{3}$ = _____ = _____ = _____        8.  $\frac{4}{5}$ = _____

4.  $\frac{2}{3}$ = _____ = _____ = _____        9.  $\frac{1}{6}$ = _____

5.  $\frac{1}{4}$ = _____ = _____                10.  $\frac{5}{6}$ = _____

6.  $\frac{3}{4}$ = _____ = _____                11.  $\frac{3}{9}$ = _____

7.  $\frac{2}{5}$ = _____                        12.  $\frac{6}{9}$ = _____

13. Look at the sets of fractions you named.  Do you see any patterns?

    Describe them: _____

    _____

14. Can you think of a rule for finding equivalent fractions?  Explain it. _____

    _____

15. Prove your rule by writing an equivalent fraction for $\frac{3}{5}$ = _____ .

    Check your answer by using your fraction circle pieces.

# Lowest Terms

You know that one fraction can be named in many different ways.
For example, $\frac{6}{12}$ can be called $\frac{2}{4}$ or $\frac{3}{6}$ or $\frac{1}{2}$.

Sometimes it is easier to understand fractions when they are named in the simplest possible way. You can probably "see" $\frac{1}{2}$ in your mind more easily than $\frac{2}{4}$ or $\frac{6}{12}$. When you rename a fraction in the simplest possible way, you are finding the *lowest terms* for that fraction.

*Try this:*

• Show $\frac{4}{8}$ on the unit circle below with your fraction circle pieces.

• Find all the equivalent fractions for $\frac{4}{8}$ that you can.
  (Remember to use pieces of *one color.*)

• Record all the equivalent fractions you find on the chart.

• Circle the equivalent fraction that used the *fewest* pieces to cover $\frac{4}{8}$.

  This is the *lowest term* for $\frac{4}{8}$.

Follow the steps above to find equivalent fractions and the *lowest term* for each fraction on the chart. Circle each lowest term.

| Fraction Name | Equivalent Fractions |
|---|---|
| $\frac{4}{8}$ | |
| $\frac{6}{9}$ | |
| $\frac{9}{12}$ | |
| $\frac{9}{10}$ | |
| $\frac{2}{6}$ | |

# Fractions Larger Than 1

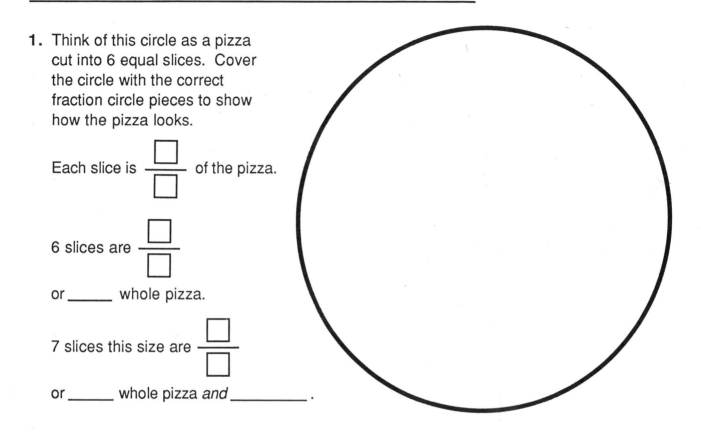

1. Think of this circle as a pizza cut into 6 equal slices. Cover the circle with the correct fraction circle pieces to show how the pizza looks.

   Each slice is $\dfrac{\square}{\square}$ of the pizza.

   6 slices are $\dfrac{\square}{\square}$

   or _____ whole pizza.

   7 slices this size are $\dfrac{\square}{\square}$

   or _____ whole pizza *and* _____ .

2. On the chart below, write the number of pizza slices as a fraction. Then tell how many pizzas and parts of a pizza it is. Use your fraction circle pieces to help you.

| Number of Slices | Fraction | Whole Pizza and Parts of a Pizza |
|:---:|:---:|:---:|
| 7 | $\dfrac{7}{6}$ | 1  and  $\dfrac{1}{6}$  =  $\dfrac{7}{6}$ |
| 8 | | and  = |
| 10 | | and  = |
| 11 | | and  = |
| 13 | | and  = |

3. Prove that $\dfrac{9}{6}$ is another name for $1\dfrac{3}{6}$.

   Prove that $\dfrac{12}{6}$ is another name for 2.

# Pizza Puzzles

1. Use your fraction circle pieces and a unit circle to help you complete and prove these problems.

   If a pizza was cut into 8 equal slices, each slice would = $\dfrac{\square}{\square}$

   8 slices = $\dfrac{\square}{\square}$ or

   _____ whole pizza.

   9 slices = $\dfrac{\square}{\square}$ or

   _____ whole pizza and $\dfrac{\square}{\square}$

   11 slices = $\dfrac{\square}{\square}$ or _____ .

   14 slices = $\dfrac{\square}{\square}$ or _____ .

2. Use your fraction circle pieces and a unit circle to cut the pizza into 10 equal slices.

   1 slice = _____

   5 slices = _____ or _____

   10 slices = _____ or _____

   12 slices = _____ or _____

3. Cut the pizza into 12 equal slices.

   1 slice = _____

   5 slices = _____

   8 slices = _____ or _____

   10 slices = _____ or _____

   12 slices = _____ or _____

   16 slices = _____ or _____

   20 slices = _____ or _____

   24 slices = _____ or _____

Write the symbol >, <, or = between each pair of numbers to make a true sentence.

4. $\boxed{1\dfrac{2}{10} \qquad \dfrac{13}{10}}$

5. $\boxed{1\dfrac{7}{8} \qquad \dfrac{15}{8}}$

6. $\boxed{2\dfrac{1}{2} \qquad \dfrac{6}{2}}$

7. $\boxed{2\dfrac{2}{3} \qquad \dfrac{8}{3}}$

# Addition of Fractions

**1.** Cover $\frac{1}{2}$ *exactly* with pieces of one color.

Do it again with another color.
Record your findings on the chart below.
Then write a fraction addition sentence
to show your discovery.

Find at least three different ways to
write addition sentences for $\frac{1}{2}$
using pieces of one color.

| Fraction | Number of Pieces to Cover It | Color | Fraction Name | Addition Sentence |
|---|---|---|---|---|
| $\frac{1}{2}$ | | | | $\frac{1}{2} =$ |
| $\frac{1}{2}$ | | | | $\frac{1}{2} =$ |
| $\frac{1}{2}$ | | | | $\frac{1}{2} =$ |
| $\frac{1}{3}$ | | | | $\frac{1}{3} =$ |
| $\frac{1}{3}$ | | | | $\frac{1}{3} =$ |

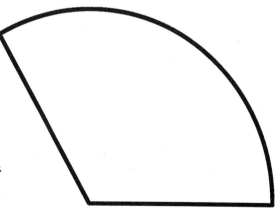

**2.** Cover $\frac{1}{3}$ *exactly* with pieces of one color.

Try to do the same thing with another color.
Find at least two different addition sentences for $\frac{1}{3}$.
Record your findings on the chart above.

# Two Colors to Add

Cover $\frac{1}{2}$ *exactly* with pieces of two different colors. Do this as many ways as you can. Record your findings on the chart below and write the corresponding addition sentence.

*For example:* $\frac{1}{2}$ can be covered by ...

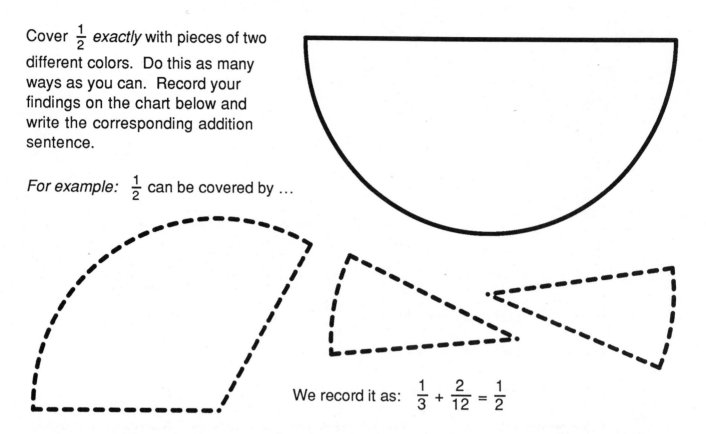

We record it as: $\frac{1}{3} + \frac{2}{12} = \frac{1}{2}$

| First Color | Number of Pieces | Fraction | Second Color | Number of Pieces | Fraction | Addition Sentence |
|---|---|---|---|---|---|---|
| | 1 | $\frac{1}{3}$ | | 2 | $\frac{2}{12}$ | $\frac{1}{3} + \frac{2}{12} = \frac{1}{2}$ |
| | | | | | | $= \frac{1}{2}$ |
| | | | | | | $= \frac{1}{2}$ |
| | | | | | | $= \frac{1}{2}$ |
| | | | | | | $= \frac{1}{2}$ |
| | | | | | | $= \frac{1}{2}$ |
| | | | | | | $= \frac{1}{2}$ |
| | | | | | | $= \frac{1}{2}$ |
| | | | | | | $= \frac{1}{2}$ |
| | | | | | | $= \frac{1}{2}$ |
| | | | | | | $= \frac{1}{2}$ |

# Adding to Make $\frac{3}{4}$

Cover $\frac{3}{4}$ *exactly* with pieces of two different colors. Do this as many ways as you can. Record your findings on the chart below and write the corresponding addition sentence.

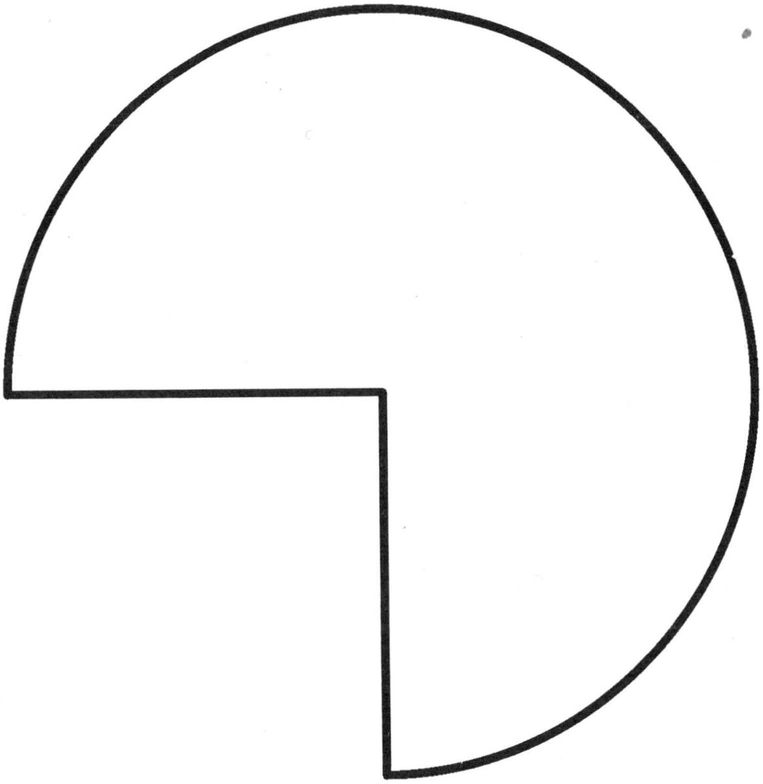

| First Color | Number of Pieces | Fraction | Second Color | Number of Pieces | Fraction | Addition Sentence |
|---|---|---|---|---|---|---|
| | | | | | | $= \frac{3}{4}$ |
| | | | | | | $= \frac{3}{4}$ |
| | | | | | | $= \frac{3}{4}$ |
| | | | | | | $= \frac{3}{4}$ |
| | | | | | | $= \frac{3}{4}$ |
| | | | | | | $= \frac{3}{4}$ |
| | | | | | | $= \frac{3}{4}$ |
| | | | | | | $= \frac{3}{4}$ |
| | | | | | | $= \frac{3}{4}$ |
| | | | | | | $= \frac{3}{4}$ |

# Two Color Addition

Cover $\frac{5}{6}$ *exactly* with pieces of two different colors. Do this as many ways as you can. Record your findings on the chart below and write the corresponding addition sentence.

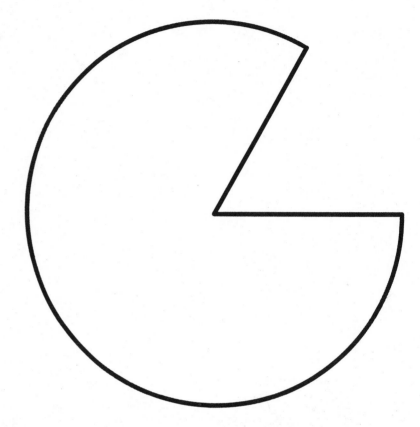

| First Color | Number of Pieces | Fraction | Second Color | Number of Pieces | Fraction | Addition Sentence |
|---|---|---|---|---|---|---|
| | | | | | | $= \frac{5}{6}$ |
| | | | | | | $= \frac{5}{6}$ |
| | | | | | | $= \frac{5}{6}$ |
| | | | | | | $= \frac{5}{6}$ |
| | | | | | | $= \frac{5}{6}$ |
| | | | | | | $= \frac{5}{6}$ |
| | | | | | | $= \frac{5}{6}$ |
| | | | | | | $= \frac{5}{6}$ |
| | | | | | | $= \frac{5}{6}$ |
| | | | | | | $= \frac{5}{6}$ |

# What's Missing?

Use your fraction circle pieces and the unit circles.
Write a fraction that completes each addition
sentence.

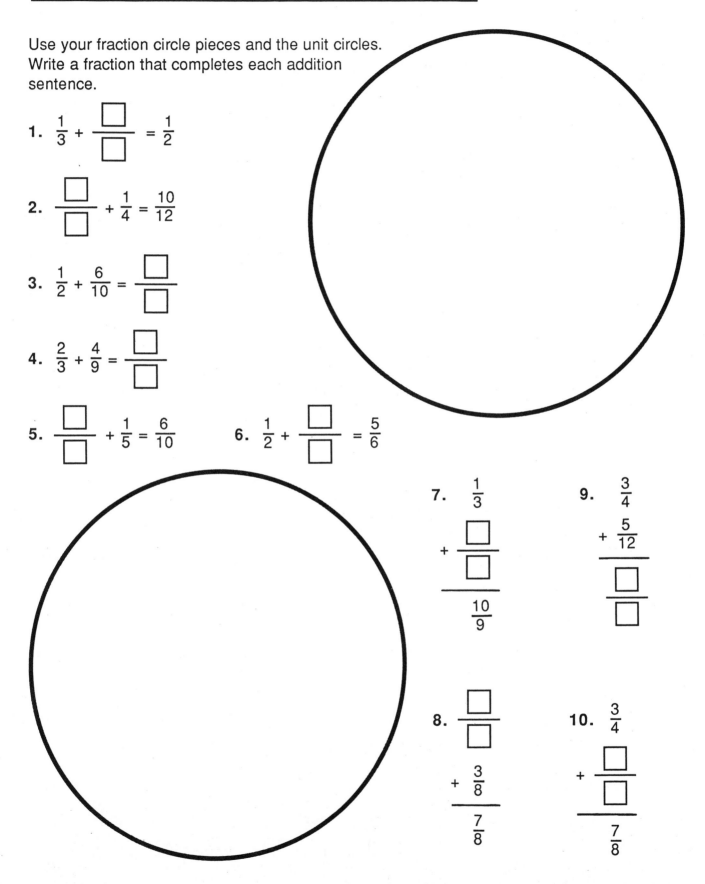

1. $\dfrac{1}{3} + \dfrac{\square}{\square} = \dfrac{1}{2}$

2. $\dfrac{\square}{\square} + \dfrac{1}{4} = \dfrac{10}{12}$

3. $\dfrac{1}{2} + \dfrac{6}{10} = \dfrac{\square}{\square}$

4. $\dfrac{2}{3} + \dfrac{4}{9} = \dfrac{\square}{\square}$

5. $\dfrac{\square}{\square} + \dfrac{1}{5} = \dfrac{6}{10}$

6. $\dfrac{1}{2} + \dfrac{\square}{\square} = \dfrac{5}{6}$

7. $\dfrac{1}{3}$
   $+ \dfrac{\square}{\square}$
   _____
   $\dfrac{10}{9}$

9. $\dfrac{3}{4}$
   $+ \dfrac{5}{12}$
   _____
   $\dfrac{\square}{\square}$

8. $\dfrac{\square}{\square}$
   $+ \dfrac{3}{8}$
   _____
   $\dfrac{7}{8}$

10. $\dfrac{3}{4}$
    $+ \dfrac{\square}{\square}$
    _____
    $\dfrac{7}{8}$

# Now What's Missing?

1. Use the following steps and your fraction circle pieces to find what's missing in the problem:

$$\frac{3}{4} = \frac{1}{2} + \frac{\Box}{\Box}$$

STEP 1. Place three $\frac{1}{4}$ pieces in the unit circle. Trace around them, label the outline $\frac{3}{4}$, and color it to match your fraction circle pieces.

STEP 2. Place a $\frac{1}{2}$ piece on top of the section you colored.

STEP 3. Find one (or more) pieces of one color to fill in the rest of the $\frac{3}{4}$ section. What pieces did you use?

_____

Fill in what's missing in the problem above.

2. Try to find other ways.

$$\frac{3}{4} = \frac{1}{2} + \frac{\Box}{\Box} \qquad \frac{3}{4} = \frac{1}{2} + \frac{\Box}{\Box}$$

3. Follow the same steps to solve this problem:

$$\frac{1}{2} + \frac{\Box}{\Box} = \frac{9}{10}$$

4. Which pieces did you use to trace the $\frac{9}{10}$ outline? _____

_____

5. Find at least two different ways to solve the problem:

$$\frac{1}{2} + \frac{\Box}{\Box} = \frac{9}{10} \qquad \frac{1}{2} + \frac{\Box}{\Box} = \frac{9}{10}$$

# Find What's Missing

1. Use the steps from page 30 and your fraction circle pieces to find what is missing.

$$\frac{2}{3} + \frac{\Box}{\Box} = \frac{3}{4}$$

STEP 1. Place pieces that equal $\frac{3}{4}$ in the unit circle. Trace around and color the outline to match. Label the section $\frac{3}{4}$.

STEP 2. Place pieces that equal $\frac{2}{3}$ on top of the $\frac{3}{4}$ outline.

STEP 3. Find the missing piece(s) of one color. What pieces did you use?

_____

Fill in what's missing in the problem above.

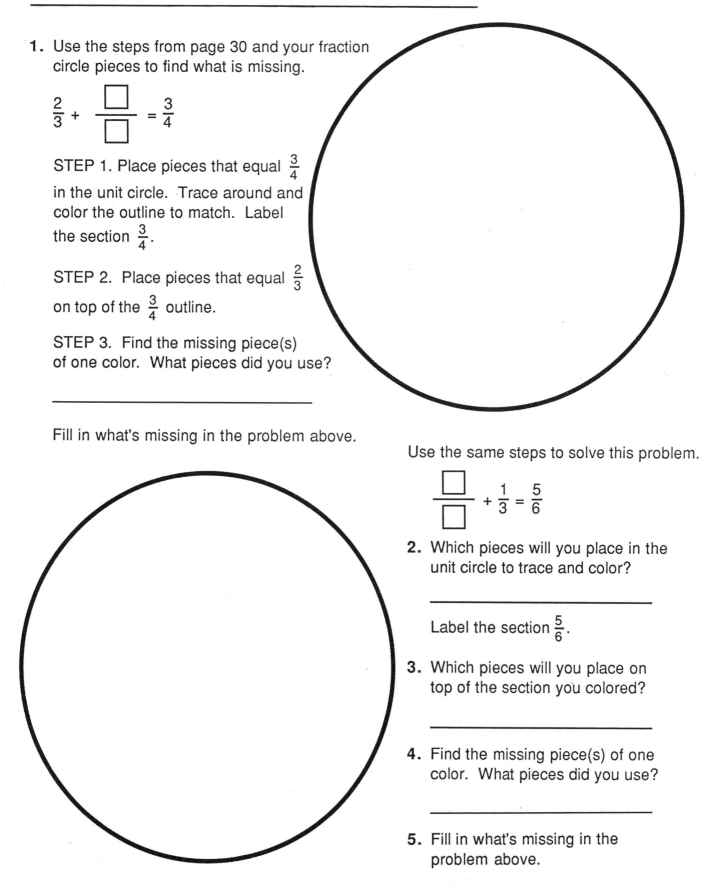

Use the same steps to solve this problem.

$$\frac{\Box}{\Box} + \frac{1}{3} = \frac{5}{6}$$

2. Which pieces will you place in the unit circle to trace and color?

_____

Label the section $\frac{5}{6}$.

3. Which pieces will you place on top of the section you colored?

_____

4. Find the missing piece(s) of one color. What pieces did you use?

_____

5. Fill in what's missing in the problem above.

# Challenge: What's Missing?

Use your fraction circle pieces to find what is missing in each problem by following the same steps you used on page 30.

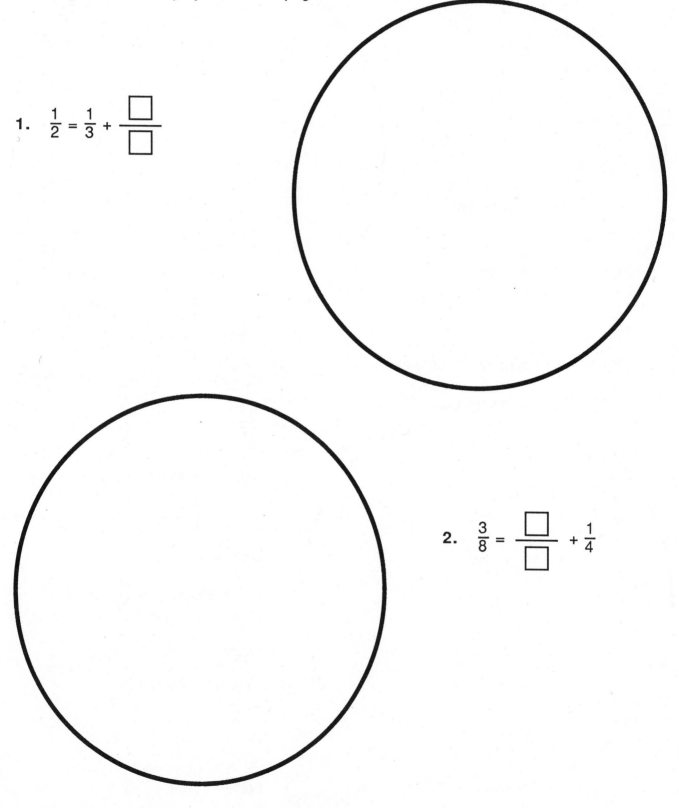

1.  $\dfrac{1}{2} = \dfrac{1}{3} + \dfrac{\square}{\square}$

2.  $\dfrac{3}{8} = \dfrac{\square}{\square} + \dfrac{1}{4}$

# Diagram What's Missing

Use your fraction circle pieces to find the missing fraction in each problem.

Use the unit circle as your workspace.

Draw a diagram of your work for each problem.

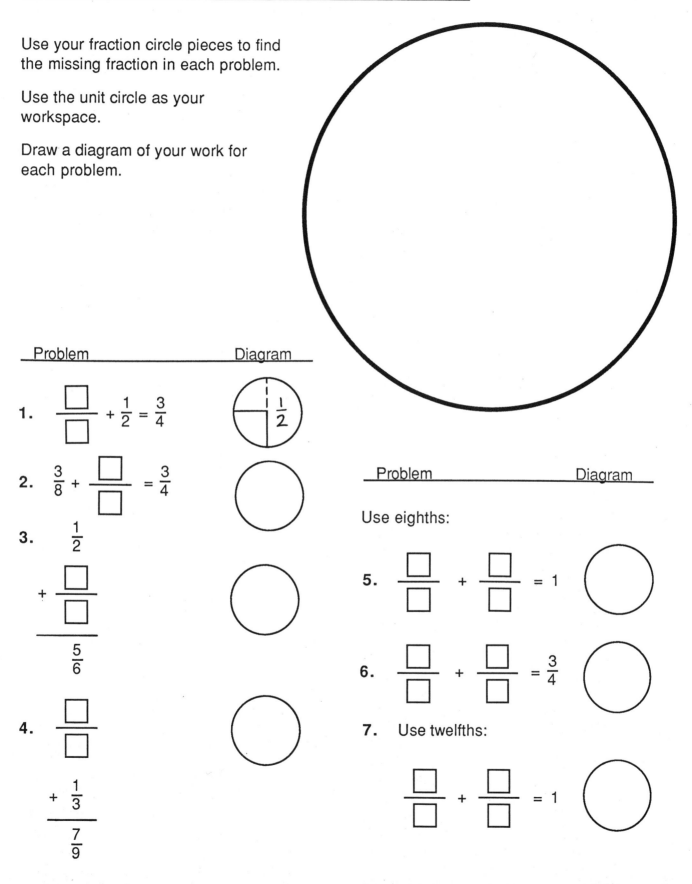

Problem                           Diagram

1. $\dfrac{\Box}{\Box} + \dfrac{1}{2} = \dfrac{3}{4}$

2. $\dfrac{3}{8} + \dfrac{\Box}{\Box} = \dfrac{3}{4}$

3. $\begin{array}{r} \dfrac{1}{2} \\ + \dfrac{\Box}{\Box} \\ \hline \dfrac{5}{6} \end{array}$

4. $\begin{array}{r} \dfrac{\Box}{\Box} \\ + \dfrac{1}{3} \\ \hline \dfrac{7}{9} \end{array}$

Problem                           Diagram

Use eighths:

5. $\dfrac{\Box}{\Box} + \dfrac{\Box}{\Box} = 1$

6. $\dfrac{\Box}{\Box} + \dfrac{\Box}{\Box} = \dfrac{3}{4}$

7. Use twelfths:

$\dfrac{\Box}{\Box} + \dfrac{\Box}{\Box} = 1$

# What's the Difference?

We can use the fraction circle pieces to find the difference for this problem.

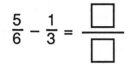

$$\frac{5}{6} - \frac{1}{3} = \frac{\Box}{\Box}$$

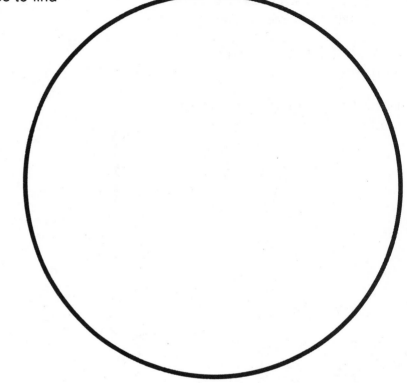

1. Which fraction is larger? Circle your answer.

   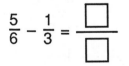

   $$\frac{5}{6} \quad \text{or} \quad \frac{1}{3}$$

2. Place the pieces for the larger fraction on the unit circle. Trace around them.

3. Place the piece for the smaller fraction on top of the outline.

4. Find the difference between the two fractions by finding a piece or pieces of *one* color to fill in the missing section.

5. The difference is:

   $$\frac{5}{6} - \frac{1}{3} = \frac{\Box}{\Box}$$

6. A diagram for this problem would look like this:

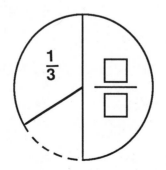

# Find the Difference

Use your fraction circle pieces to find
the difference for this problem.

$$\begin{array}{r} \dfrac{3}{4} \\[6pt] -\ \dfrac{3}{8} \\ \hline \end{array}$$

1. Decide which fraction is larger.
   Circle your answer.

   $\dfrac{3}{4}$   or   $\dfrac{3}{8}$

2. Place the pieces for the larger
   fraction on the unit circle.
   Trace around them.

3. Place the pieces for the smaller
   fraction on top of the outline.

4. Find the *difference* between the two
   fractions by finding a piece or pieces of
   one color to fill in the missing section.

   The difference is:

   $\dfrac{3}{4} - \dfrac{3}{8} = \dfrac{\square}{\square}$

5. A diagram for this problem would look like this:

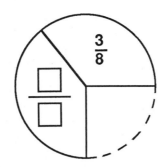

# Now What's the Difference?

Use your fraction circle pieces to find the difference.

$$\frac{7}{12} - \frac{1}{4} = \frac{\square}{\square}$$

1. The larger fraction is _____ .

2. Place the pieces for the larger fraction in the unit circle and trace around them.

3. Place the smaller fraction on top.

4. Find the difference.

$$\frac{7}{12} - \frac{1}{4} = \frac{\square}{\square}$$

5. Diagram:

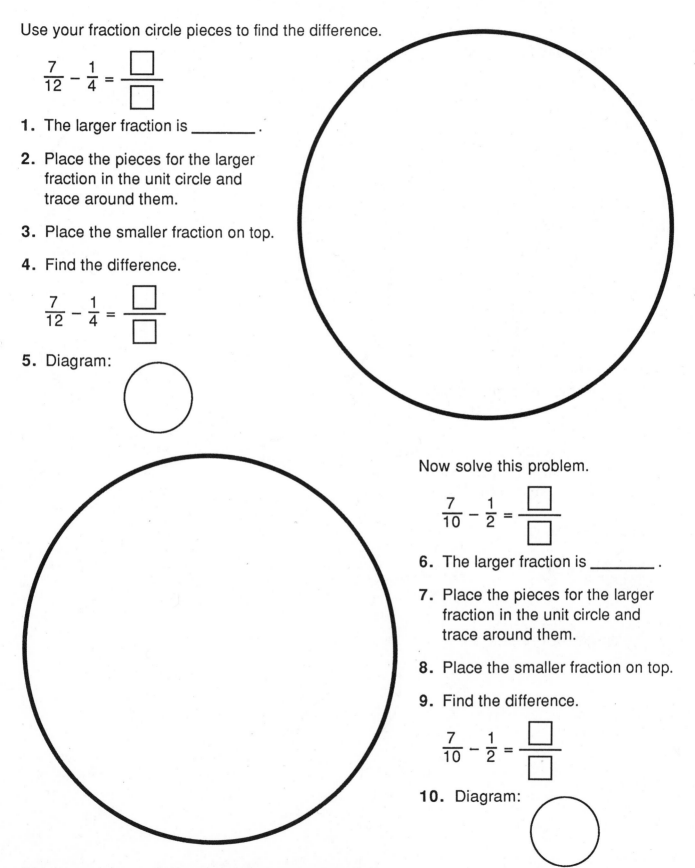

Now solve this problem.

$$\frac{7}{10} - \frac{1}{2} = \frac{\square}{\square}$$

6. The larger fraction is _____ .

7. Place the pieces for the larger fraction in the unit circle and trace around them.

8. Place the smaller fraction on top.

9. Find the difference.

$$\frac{7}{10} - \frac{1}{2} = \frac{\square}{\square}$$

10. Diagram:

# Finding Differences

Use your fraction circle pieces to find the difference in each problem.

Use the unit circle as your workspace.

Draw a diagram of your work for each problem.

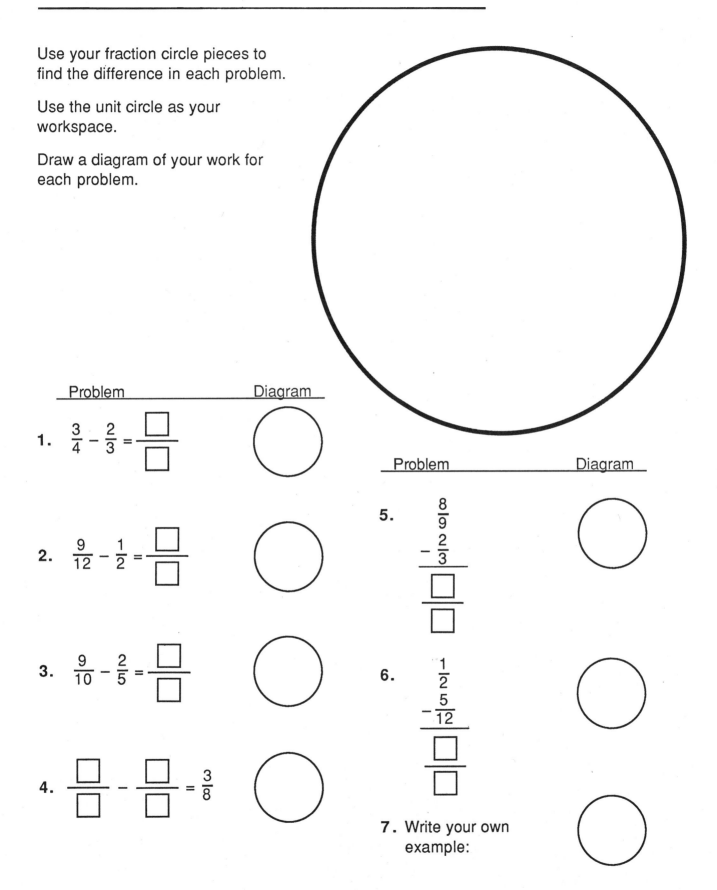

| Problem | Diagram |
|---|---|
| 1. $\dfrac{3}{4} - \dfrac{2}{3} = \dfrac{\square}{\square}$ | |
| 2. $\dfrac{9}{12} - \dfrac{1}{2} = \dfrac{\square}{\square}$ | |
| 3. $\dfrac{9}{10} - \dfrac{2}{5} = \dfrac{\square}{\square}$ | |
| 4. $\dfrac{\square}{\square} - \dfrac{\square}{\square} = \dfrac{3}{8}$ | |

| Problem | Diagram |
|---|---|
| 5. $\begin{array}{r} \frac{8}{9} \\ -\frac{2}{3} \\ \hline \frac{\square}{\square} \end{array}$ | |
| 6. $\begin{array}{r} \frac{1}{2} \\ -\frac{5}{12} \\ \hline \frac{\square}{\square} \end{array}$ | |
| 7. Write your own example: | |

# Take Away

The example $\frac{2}{3} - \frac{1}{3} = \frac{\Box}{\Box}$ can mean "If we have $\frac{2}{3}$ and we take away $\frac{1}{3}$,

what is left?" Use your fraction circle pieces to find the answer.

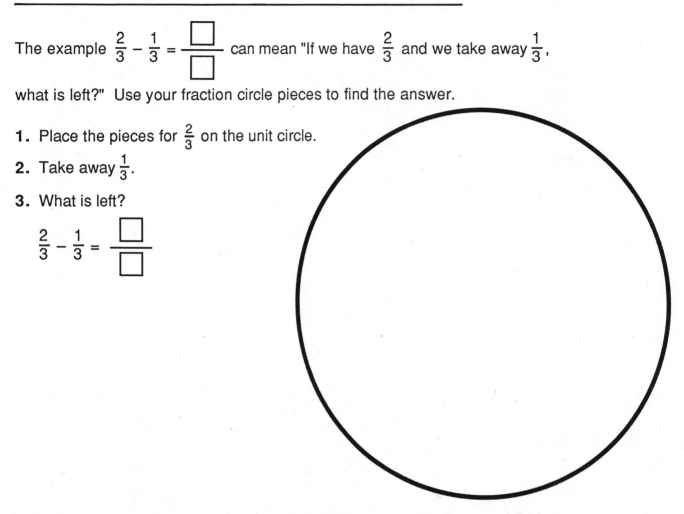

**1.** Place the pieces for $\frac{2}{3}$ on the unit circle.

**2.** Take away $\frac{1}{3}$.

**3.** What is left?

$$\frac{2}{3} - \frac{1}{3} = \frac{\Box}{\Box}$$

In the problems below, use your fraction circle pieces to take away and find what is left. Use the unit circle above as your workspace.

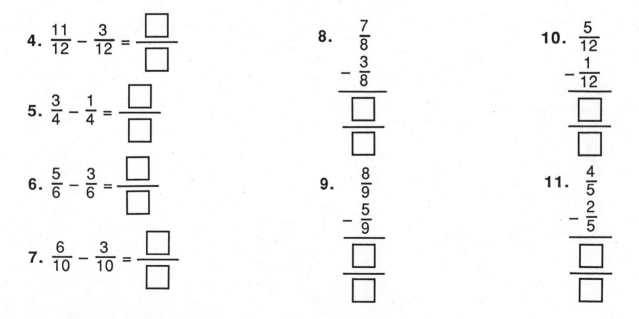

**4.** $\frac{11}{12} - \frac{3}{12} = \frac{\Box}{\Box}$

**5.** $\frac{3}{4} - \frac{1}{4} = \frac{\Box}{\Box}$

**6.** $\frac{5}{6} - \frac{3}{6} = \frac{\Box}{\Box}$

**7.** $\frac{6}{10} - \frac{3}{10} = \frac{\Box}{\Box}$

**8.** $\begin{array}{r} \frac{7}{8} \\ -\frac{3}{8} \\ \hline \frac{\Box}{\Box} \end{array}$

**9.** $\begin{array}{r} \frac{8}{9} \\ -\frac{5}{9} \\ \hline \frac{\Box}{\Box} \end{array}$

**10.** $\begin{array}{r} \frac{5}{12} \\ -\frac{1}{12} \\ \hline \frac{\Box}{\Box} \end{array}$

**11.** $\begin{array}{r} \frac{4}{5} \\ -\frac{2}{5} \\ \hline \frac{\Box}{\Box} \end{array}$

# One Color Take Away

It is easy to take away and find out what is left when two fraction circle pieces are the same color. They belong to the same family and have the same denominator. This is called a *common denominator.*

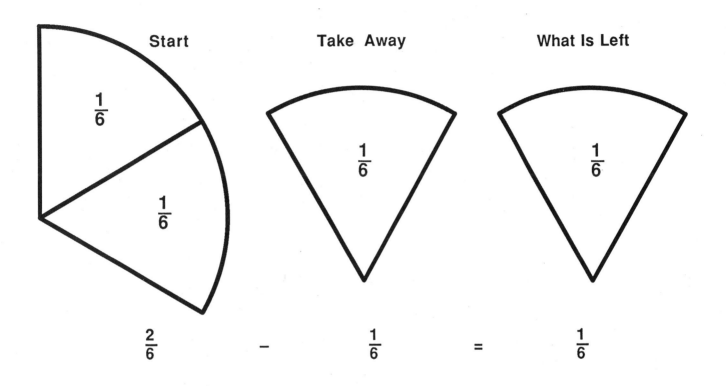

Use your fraction circle pieces to help you solve these problems.

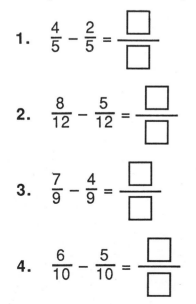

1. $\dfrac{4}{5} - \dfrac{2}{5} = \dfrac{\square}{\square}$

2. $\dfrac{8}{12} - \dfrac{5}{12} = \dfrac{\square}{\square}$

3. $\dfrac{7}{9} - \dfrac{4}{9} = \dfrac{\square}{\square}$

4. $\dfrac{6}{10} - \dfrac{5}{10} = \dfrac{\square}{\square}$

# Common Denominators

When two fraction circle pieces have different
colors, the denominators are different, and it is
hard to know how much to take away.

$$\begin{array}{r} \frac{1}{4} \\ -\frac{1}{6} \\ \hline \end{array}$$

You can make this problem easier by finding a common color to cover both
fractions before you take away. **This common color is the *common
denominator* for the two fractions.** Then rename each fraction as an
equivalent fraction using the common denominator.

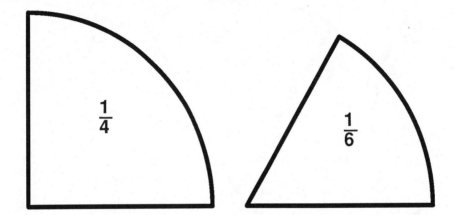

1. A common color for pieces that can cover $\frac{1}{4}$ and $\frac{1}{6}$ *exactly* is _____.

2. Each piece of this color is called a _____.

3. Cover $\frac{1}{4}$ and $\frac{1}{6}$ with pieces of the same color.

4. It takes _____ pieces to cover $\frac{1}{4}$ exactly.
   Write this as an equivalent fraction.      $\frac{1}{4} = \frac{\square}{\square}$

5. It takes _____ pieces to cover $\frac{1}{6}$ exactly.
   Write this an as equivalent fraction.      $\frac{1}{6} = \frac{\square}{\square}$

6. Now you have renamed both fractions as
   equivalents with *common denominators.*

   Rewrite the example using the equivalent
   fractions you just found. Then solve the
   problem using your fraction circle pieces.

   $$\begin{array}{r} \frac{1}{4} = \\ -\frac{1}{6} = \\ \hline \end{array}$$

# Add and Subtract

We need to find common denominators in order to add or subtract fractions that have different denominators.

For the fractions below, use your fraction circle pieces to:

- Find pieces with a common color to cover both fractions.
- Name the common denominator.
- Rename each fraction as an equivalent fraction using the common denominator.
- Solve the problems.

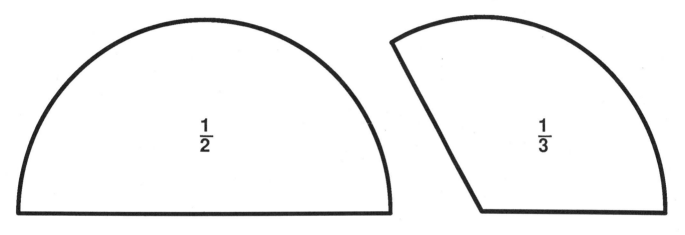

$\frac{1}{2}$ and $\frac{1}{3}$

**1.** Common color: _____

**2.** Common denominator: _____

**3.** Equivalent fractions using the common denominator:

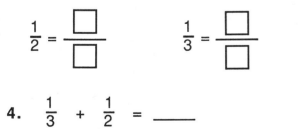

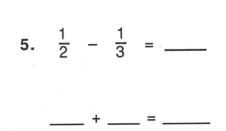

# Add and Subtract Again

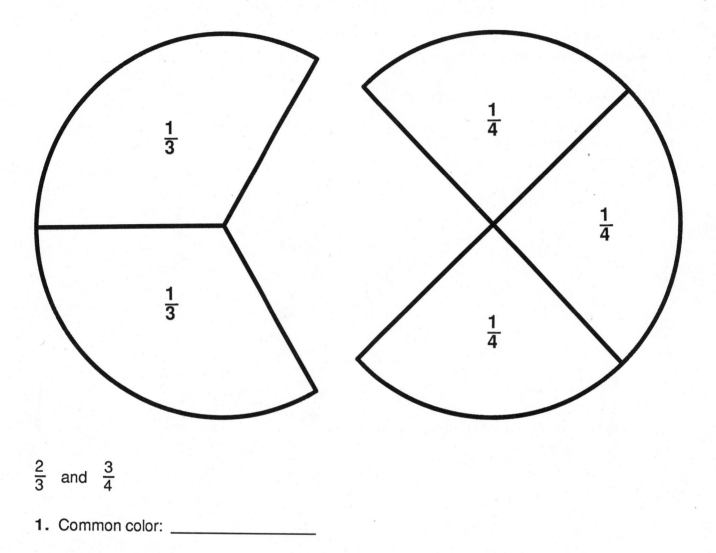

$\dfrac{2}{3}$  and  $\dfrac{3}{4}$

**1.** Common color: _____

**2.** Common denominator: _____

**3.** Equivalent fractions using the common denominator:

$\dfrac{2}{3} = \dfrac{\square}{\square}$                    $\dfrac{3}{4} = \dfrac{\square}{\square}$

Use your fraction circle pieces to solve these problems.

**4.**  $\dfrac{2}{3}$  =

$+\dfrac{3}{4}$  =

_____

**5.**  $\dfrac{3}{4}$  =

$-\dfrac{2}{3}$  =

_____

# Addition and Subtraction

Sometimes one of the fractions *already has* the common denominator.  Use your fraction circle pieces to:

• Find a common color to cover both fractions.
• Name the common denominator.
• Rename each fraction as an equivalent fraction using the common denominator.
• Solve.  Use the unit circles as your workspace.

1. $\dfrac{2}{3}$ $-$ $\dfrac{1}{6}$ = _____

\_\_\_\_\_ $-$ \_\_\_\_\_ = \_\_\_\_\_

2. $\dfrac{1}{3}$ =

$+\dfrac{2}{9}$ =
_____

3. $\dfrac{4}{6}$ $+$ $\dfrac{2}{12}$ = \_\_\_\_\_

\_\_\_\_\_ $+$ \_\_\_\_\_ = \_\_\_\_\_

4. $\dfrac{7}{10}$ =

$-\dfrac{1}{2}$ =
_____

5. $\dfrac{11}{12}$ $-$ $\dfrac{3}{4}$ = \_\_\_\_\_

\_\_\_\_\_ $-$ \_\_\_\_\_ = \_\_\_\_\_

6. $\dfrac{6}{8}$ $+$ $\dfrac{3}{4}$ = \_\_\_\_\_

\_\_\_\_\_ $+$ \_\_\_\_\_ = \_\_\_\_\_

# Practice Adding and Subtracting

Use your fraction circle pieces to:

• Find a common color to cover both fractions.
• Name the common denominator.
• Rename each fraction as an equivalent fraction using the common denominator.
• Solve. Use the unit circles as your workspace.

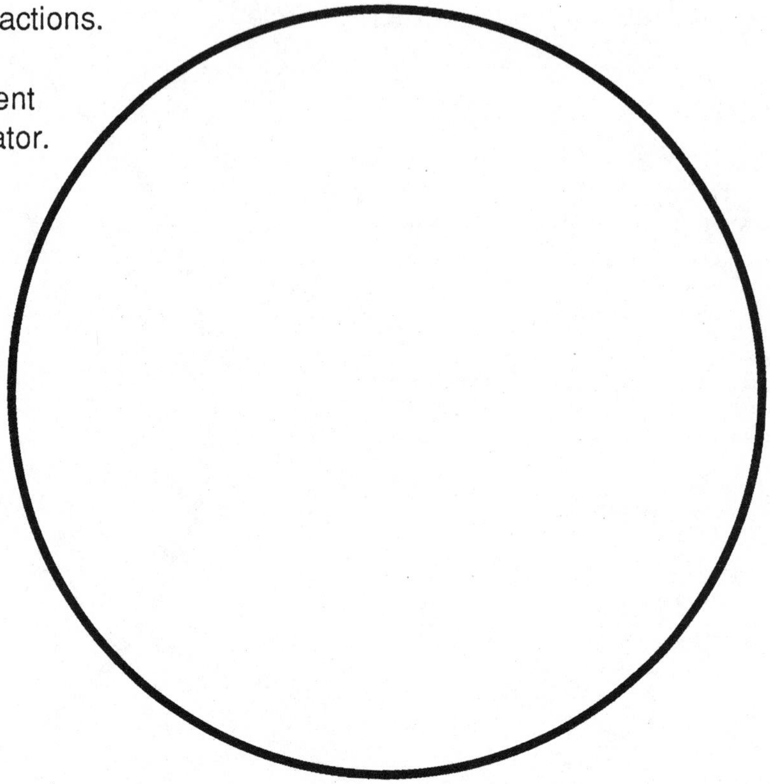

1.   $\frac{1}{2}$ =

   $+\frac{2}{5}$ =
   _____

2.   $\frac{5}{10}$ =

   $+\frac{3}{6}$ =
   _____

3.   $\frac{4}{8}$ =

   $+\frac{3}{12}$ =
   _____

4.   $\frac{2}{3}$ =

   $+\frac{5}{9}$ =
   _____

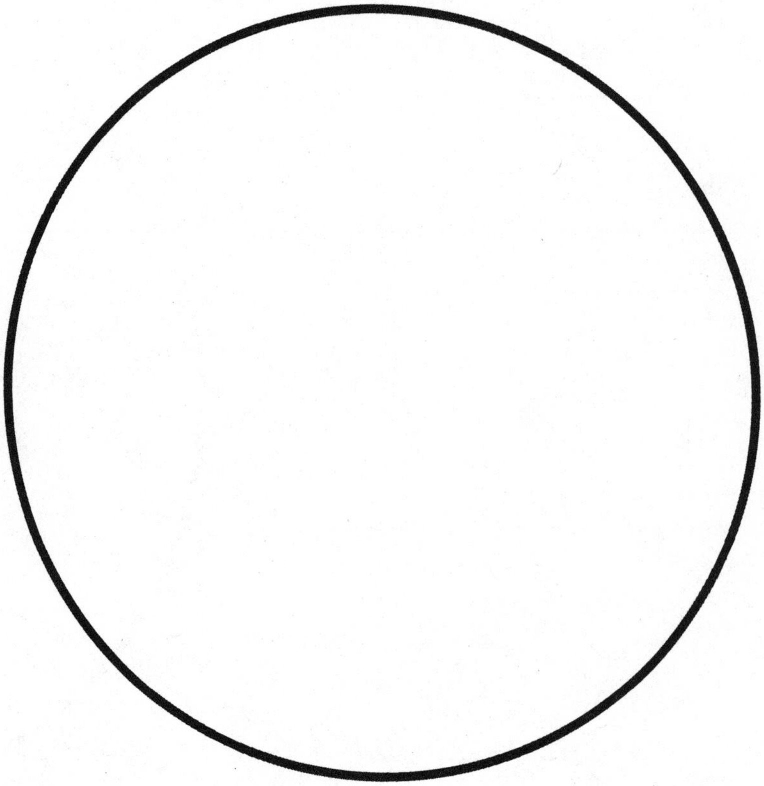

5. $\frac{3}{5}$  –  $\frac{1}{2}$ = _____

   ____ – ____ = ____

6. $\frac{1}{3}$  +  $\frac{8}{12}$ = _____

   ____ + ____ = ____

# Fraction Challenge

Rename each fraction as an equivalent fraction using a common denominator. Solve each problem, using the unit circles as your workspace.

**1.**  $\dfrac{2}{6}$ =

$+\dfrac{6}{9}$ =

_____

**2.**  $\dfrac{8}{12}$ – $\dfrac{3}{9}$ = _____

_____ – _____ = _____

**3.**  $1\dfrac{1}{4}$ =

$+\dfrac{2}{3}$ =

_____

**4.**  $\dfrac{3}{4}$ =

$-\dfrac{2}{6}$ =

_____

**5.** $\dfrac{3}{6}$ + $\dfrac{4}{8}$ = _____

_____ + _____ = _____

**6.**  $\dfrac{9}{6}$ =

$+\dfrac{1}{2}$ =

_____

# Answer Key

**Page 1:**

White. (This is the color of the unit circle given on the back cover foldout of this book. The color will vary if you had your students color the unit circle a different color.)

**Page 2:**

1. The color of the $\frac{1}{2}$ piece will vary depending on the set of fraction circle pieces you use; yes.

2. 2
3. two

**Page 3:**

1. The color of the $\frac{1}{3}$ piece will vary depending on the set of fraction circle pieces you use; yes.

2. 3; 3

3. 3; $\frac{1}{3}$

4. $\frac{2}{3}$; $\frac{3}{3}$

5. 1 out of 3; $\frac{1}{3}$

**Page 4:**

1. The color of the $\frac{1}{4}$ piece will vary depending on the set of fraction circle pieces you use; yes.

2. 4; 4

3. 4; $\frac{1}{4}$

4. $\frac{2}{4}$; $\frac{3}{4}$; $\frac{4}{4}$

5. 1 out of 4; $\frac{1}{4}$

**Page 5:**

1. Pink is the color of the $\frac{1}{5}$ piece on the back cover foldout of this book; yes.

2. 5

3. $\frac{1}{5}$; one-fifth; answers will vary, but should be similar to, "It is one of five pieces that are needed to cover the unit circle."

4. $\frac{2}{5}$; $\frac{3}{5}$; $\frac{4}{5}$; $\frac{5}{5}$

5. 1 out of 5; $\frac{1}{5}$

**Page 6:**

1.  The color of the $\frac{1}{6}$ piece will vary depending on the set of fraction circle pieces you use; yes.

2.  6

3.  $\frac{1}{6}$; one-sixth; answers will vary, but should be similar to, "It is one of six pieces needed to cover the unit circle."

4.  $\frac{2}{6}$; $\frac{3}{6}$; $\frac{4}{6}$; $\frac{5}{6}$; $\frac{6}{6}$

5.  1 out of 6; $\frac{1}{6}$

**Page 7:**

1.  The color of the $\frac{1}{8}$ piece will vary depending on the set of fraction circle pieces you use; yes.

2.  8

3.  $\frac{1}{8}$; one-eighth; answers will vary, but should be similar to, "It is one of eight pieces needed to cover the unit circle."

4.  $\frac{2}{8}$; $\frac{3}{8}$; $\frac{4}{8}$; $\frac{5}{8}$; $\frac{6}{8}$; $\frac{7}{8}$; $\frac{8}{8}$

5.  1 out of 8; $\frac{1}{8}$

**Page 8:**

1.  The color of the $\frac{1}{9}$ piece will vary depending on the set of fraction circle pieces you use; yes.

2.  9

3.  $\frac{1}{9}$; one-ninth; answers will vary, but should be similar to, "It is one of nine pieces needed to cover the unit circle."

4.  $\frac{2}{9}$; $\frac{3}{9}$; $\frac{4}{9}$; $\frac{5}{9}$; $\frac{6}{9}$; $\frac{7}{9}$; $\frac{8}{9}$; $\frac{9}{9}$

5.  1 out of 9; $\frac{1}{9}$

**Page 9:**

1.  Purple is the color of the $\frac{1}{10}$ pieces on the back cover foldout of this book; yes.

2.  10

3.  $\frac{1}{10}$; one-tenth; answers will vary, but should be similar to, "It is one of ten pieces needed to cover the unit circle."

4.  $\frac{2}{10}$; $\frac{3}{10}$; $\frac{4}{10}$; $\frac{5}{10}$; $\frac{6}{10}$; $\frac{7}{10}$; $\frac{8}{10}$; $\frac{9}{10}$; $\frac{10}{10}$

5.  1 out of 10; $\frac{1}{10}$

**Page 10:**

1.  Orange is the color of the $\frac{1}{12}$ pieces on the back cover foldout of this book; yes.

2.  12

3.  $\frac{1}{12}$; one-twelfth; answers will vary, but should be similar to, "It is one of twelve pieces needed to cover the unit circle."

4.  $\frac{2}{12}$; $\frac{3}{12}$; $\frac{4}{12}$; $\frac{5}{12}$; $\frac{6}{12}$; $\frac{7}{12}$; $\frac{8}{12}$; $\frac{9}{12}$; $\frac{10}{12}$; $\frac{11}{12}$; $\frac{12}{12}$

5.  1 out of 12; $\frac{1}{12}$

**1.**

| | | Names for 1 | |
|---|---|---|---|
| Color * | Number of pieces to make 1 | Fraction Name | |
| | | Numbers | Words |
| | 2 | $\frac{1}{2}$ | one-half |
| | 3 | $\frac{1}{3}$ | one-third |
| | 4 | $\frac{1}{4}$ | one-fourth |
| | 5 | $\frac{1}{5}$ | one-fifth |
| | 6 | $\frac{1}{6}$ | one-sixth |
| | 8 | $\frac{1}{8}$ | one-eighth |
| | 9 | $\frac{1}{9}$ | one-ninth |
| | 10 | $\frac{1}{10}$ | one-tenth |
| | 12 | $\frac{1}{12}$ | one-twelfth |

* Note that the color of the fraction circle pieces will vary depending on the set you use.

2. Answers will vary. One possibility: To make 1, the number on top of the fraction and the number on the bottom of the fraction are both the same.

---

**Page 12:**
1. 4
2. 4 out of 4; $\frac{4}{4}$
3. $\frac{3}{4}$
4. the bottom
5. the top

**Page 13:**
Answers will vary. Sample problems include:

6 families were invited and 5 went = $\frac{5}{6}$

2 families were invited and 3 went = $\frac{2}{3}$

5 families were invited and 2 went = $\frac{2}{5}$

**Page 14:**
Each fraction circle piece is shown here:

$\frac{1}{2}, \frac{1}{3}, \frac{1}{4}, \frac{1}{5}, \frac{1}{6}, \frac{1}{8}, \frac{1}{9}, \frac{1}{10}, \frac{1}{12}$

Students should match their fraction circle pieces to the appropriate outlines.

**Page 15:**

1. Largest to smallest:

$$\frac{1}{2}, \frac{1}{3}, \frac{1}{4}, \frac{1}{5}, \frac{1}{6}, \frac{1}{8}, \frac{1}{9}, \frac{1}{10}, \frac{1}{12}$$

2. Largest: $\frac{1}{2}$; 2    Smallest: $\frac{1}{12}$; 12

   The unit circle is the same size but the fraction pieces are smaller, so it takes more pieces to fill up the circle.

3. Answers will vary, but should be similar to:

   The denominators get larger as the fractions get smaller.

   It takes more pieces to cover the unit circle as the fractions get smaller.

**Page 16:**

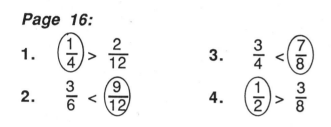

1. $\left(\frac{1}{4}\right) > \frac{2}{12}$

2. $\frac{3}{6} < \left(\frac{9}{12}\right)$

3. $\frac{3}{4} < \left(\frac{7}{8}\right)$

4. $\left(\frac{1}{2}\right) > \frac{3}{8}$

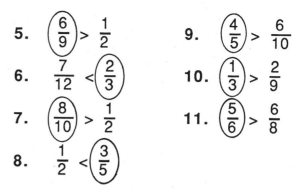

5. $\left(\frac{6}{9}\right) > \frac{1}{2}$

6. $\frac{7}{12} < \left(\frac{2}{3}\right)$

7. $\left(\frac{8}{10}\right) > \frac{1}{2}$

8. $\frac{1}{2} < \left(\frac{3}{5}\right)$

9. $\left(\frac{4}{5}\right) > \frac{6}{10}$

10. $\left(\frac{1}{3}\right) > \frac{2}{9}$

11. $\left(\frac{5}{6}\right) > \frac{6}{8}$

**Page 17:**

1. $\frac{5}{8} > \frac{6}{12}$

2. $\frac{1}{2} < \frac{8}{9}$

3. $\frac{3}{4} > \frac{1}{2}$

4. $\frac{2}{5} < \frac{7}{10}$

5. $\frac{2}{3} > \frac{5}{9}$

6. $\frac{9}{12} < \frac{7}{8}$

7. $\frac{3}{5} > \frac{1}{3}$

8. $\frac{7}{12} < \frac{9}{10}$

9. $\frac{3}{4} = \frac{9}{12}$

10. $\frac{2}{3} > \frac{3}{8}$

11. $\frac{3}{6} = \frac{4}{8}$

12. $\frac{4}{10} = \frac{2}{5}$

**Page 18:**

$\frac{1}{2}$ of the unit circle

| Fraction: $\frac{1}{2}$ | | | |
|---|---|---|---|
| Number of Pieces to Cover It | Color * | Fraction | Different Fraction Names for ◠ |
| 2 | | $\frac{1}{4}$ | $\frac{2}{4}$ |
| 3 | | $\frac{1}{6}$ | $\frac{3}{6}$ |
| 4 | | $\frac{1}{8}$ | $\frac{4}{8}$ |
| 5 | | $\frac{1}{10}$ | $\frac{5}{10}$ |
| 6 | | $\frac{1}{12}$ | $\frac{6}{12}$ |

\* Note that the color of the fraction circle pieces will vary depending on the set you use.

**Page 19:**

**1.** $\frac{1}{3}$ of the unit circle

| Fraction: $\frac{1}{3}$ | | | |
|---|---|---|---|
| Number of Pieces to Cover It | Color * | Fraction | Different Fraction Names for ⬭ |
| 2 | | $\frac{1}{6}$ | $\frac{2}{6}$ |
| 3 | | $\frac{1}{9}$ | $\frac{3}{9}$ |
| 4 | | $\frac{1}{12}$ | $\frac{4}{12}$ |

\* Note that the color of the fraction circle pieces will vary depending on the set you use.

**2.** $\frac{1}{4}$ of the unit circle

| Fraction: $\frac{1}{4}$ | | | |
|---|---|---|---|
| Number of Pieces to Cover It | Color * | Fraction | Different Fraction Names for ◁ |
| 2 | | $\frac{1}{8}$ | $\frac{2}{8}$ |
| 3 | | $\frac{1}{12}$ | $\frac{3}{12}$ |

\* Note that the color of the fraction circle pieces will vary depending on the set you use.

---

**Page 20:**

**1.** $\frac{1}{2} = \frac{2}{4}$

$= \frac{3}{6}$

$= \frac{4}{8}$

$= \frac{5}{10}$

$= \frac{6}{12}$

**2.** $\frac{2}{3} = \frac{4}{6}$

$= \frac{6}{9}$

$= \frac{8}{12}$

**3.** $\frac{3}{4} = \frac{6}{8}$

$= \frac{9}{12}$

$\frac{2}{6} = \frac{1}{3}$

$= \frac{3}{9}$

$= \frac{4}{12}$

**4.** $\frac{2}{8} = \frac{1}{4}$

$= \frac{3}{12}$

**5.** $\frac{1}{6} = \frac{2}{12}$

**6.** $\frac{5}{6} = \frac{10}{12}$

**7.** $\frac{2}{5} = \frac{4}{10}$

**8.** $\frac{6}{10} = \frac{3}{5}$

9. $\frac{4}{5} = \frac{8}{10}$

10. $\frac{2}{12} = \frac{1}{6}$

$\frac{3}{12} = \frac{1}{4}$

$\frac{10}{12} = \frac{5}{6}$

11. $\frac{12}{4} = 3$

$\frac{12}{6} = 2$

$\frac{10}{5} = 2$

6. $\frac{3}{4} = \frac{6}{8} = \frac{9}{12}$

7. $\frac{2}{5} = \frac{4}{10}$

8. $\frac{4}{5} = \frac{8}{10}$

9. $\frac{1}{6} = \frac{2}{12}$

10. $\frac{5}{6} = \frac{10}{12}$

11. $\frac{3}{9} = \frac{1}{3}$

12. $\frac{6}{9} = \frac{2}{3}$

13. The numerators and denominators are multiples of the same numbers.

14. Multiply the numerators and denominators by the same numbers.

15. $\frac{6}{10}$

**Page 21:**

1. $1 = \frac{2}{2} = \frac{3}{3} = \frac{4}{4} = \frac{5}{5} = \frac{6}{6}$

$= \frac{8}{8} = \frac{9}{9} = \frac{10}{10} = \frac{12}{12}$

2. $\frac{1}{2} = \frac{2}{4} = \frac{3}{6} = \frac{4}{8} = \frac{5}{10} = \frac{6}{12}$

3. $\frac{1}{3} = \frac{2}{6} = \frac{3}{9} = \frac{4}{12}$

4. $\frac{2}{3} = \frac{4}{6} = \frac{6}{9} = \frac{8}{12}$

5. $\frac{1}{4} = \frac{2}{8} = \frac{3}{12}$

**Page 22:**

| Fraction Name | Equivalent Fractions | | | | |
|---|---|---|---|---|---|
| $\frac{4}{8}$ | $\left(\frac{1}{2}\right)$ | $\frac{2}{4}$ | $\frac{3}{6}$ | $\frac{5}{10}$ | $\frac{6}{12}$ |
| $\frac{6}{9}$ | $\left(\frac{2}{3}\right)$ | $\frac{4}{6}$ | $\frac{8}{12}$ | | |
| $\frac{9}{12}$ | $\left(\frac{3}{4}\right)$ | $\frac{6}{8}$ | | | |
| $\frac{9}{10}$ | $\left(\frac{9}{10}\right)$ | | | | |
| $\frac{2}{6}$ | $\left(\frac{1}{3}\right)$ | $\frac{3}{9}$ | $\frac{4}{12}$ | | |

**Page 23:**

1. $\frac{1}{6}$; $\frac{6}{6}$ or 1 whole pizza; $\frac{7}{6}$ or 1 whole pizza and $\frac{1}{6}$

2.

| Number of Slices | Fraction | Whole Pizza and Parts of a Pizza |
|---|---|---|
| 7 | $\frac{7}{6}$ | 1 and $\frac{1}{6}$ = $\frac{7}{6}$ |
| 8 | $\frac{8}{6}$ | 1 and $\frac{2}{6}$ = $\frac{8}{6}$ |
| 10 | $\frac{10}{6}$ | 1 and $\frac{4}{6}$ = $\frac{10}{6}$ |
| 11 | $\frac{11}{6}$ | 1 and $\frac{5}{6}$ = $\frac{11}{6}$ |
| 13 | $\frac{13}{6}$ | 1 and $\frac{7}{6}$ = $\frac{13}{6}$ |

---

**Page 24:**

Accept all correct student answers (equivalent fractions) for the problems on this page.

1. $\frac{1}{8}$; $\frac{8}{8}$ or 1 whole pizza; $\frac{9}{8}$ or 1 whole pizza and $\frac{1}{8}$; $\frac{11}{8}$ or $1\frac{3}{8}$; $\frac{14}{8}$ or $1\frac{6}{8}$

2. 1 slice = $\frac{1}{10}$

   5 slices = $\frac{5}{10}$ (or $\frac{1}{2}$)

   10 slices = $\frac{10}{10}$ (or 1)

   12 slices = $\frac{12}{10}$ (or $1\frac{2}{10}$)

3. 1 slice = $\frac{1}{12}$

   5 slices = $\frac{5}{12}$

8 slices = $\frac{8}{12}$ (or $\frac{2}{3}$)

10 slices = $\frac{10}{12}$ (or $\frac{5}{6}$)

12 slices = $\frac{12}{12}$ (or 1)

16 slices = $\frac{16}{12}$ (or $1\frac{4}{12}$)

20 slices = $\frac{20}{12}$ (or $1\frac{8}{12}$)

24 slices = $\frac{24}{12}$ (or 2)

4. $1\frac{2}{10}$ < $\frac{13}{10}$

5. $1\frac{7}{8}$ = $\frac{15}{8}$

6. $2\frac{1}{2}$ < $\frac{6}{2}$

7. $2\frac{2}{3}$ = $\frac{8}{3}$

## *Page 25:*
All possible answers are given below.

| Fraction | Number of Pieces to Cover It | Color * | Fraction Name | Addition Sentence |
|---|---|---|---|---|
| $\frac{1}{2}$ | 2 | | $\frac{1}{4}$ | $\frac{1}{2} = \frac{1}{4} + \frac{1}{4}$ |
| $\frac{1}{2}$ | 3 | | $\frac{1}{6}$ | $\frac{1}{2} = \frac{1}{6} + \frac{1}{6} + \frac{1}{6}$ |
| $\frac{1}{2}$ | 4 | | $\frac{1}{8}$ | $\frac{1}{2} = \frac{1}{8} + \frac{1}{8} + \frac{1}{8} + \frac{1}{8}$ |
| $\frac{1}{2}$ | 5 | | $\frac{1}{10}$ | $\frac{1}{2} = \frac{1}{10} + \frac{1}{10} + \frac{1}{10} + \frac{1}{10} + \frac{1}{10}$ |
| $\frac{1}{2}$ | 6 | | $\frac{1}{12}$ | $\frac{1}{2} = \frac{1}{12} + \frac{1}{12} + \frac{1}{12} + \frac{1}{12} + \frac{1}{12} + \frac{1}{12}$ |
| $\frac{1}{3}$ | 2 | | $\frac{1}{6}$ | $\frac{1}{3} = \frac{1}{6} + \frac{1}{6}$ |
| $\frac{1}{3}$ | 3 | | $\frac{1}{9}$ | $\frac{1}{3} = \frac{1}{9} + \frac{1}{9} + \frac{1}{9}$ |
| $\frac{1}{3}$ | 4 | | $\frac{1}{12}$ | $\frac{1}{3} = \frac{1}{12} + \frac{1}{12} + \frac{1}{12} + \frac{1}{12}$ |

\* Note that the color of the fraction circle pieces will vary depending on the set you use.

***Page 26:***

Nine examples are included in the chart below. Accept any other correct student answers.

| First Color * | Number of Pieces | Fraction | Second Color * | Number of Pieces | Fraction | Addition Sentence |
|---|---|---|---|---|---|---|
| | 1 | $\frac{1}{3}$ | | 2 | $\frac{2}{12}$ | $\frac{1}{3} + \frac{2}{12} = \frac{1}{2}$ |
| | 1 | $\frac{1}{4}$ | | 2 | $\frac{2}{8}$ | $\frac{1}{4} + \frac{2}{8} = \frac{1}{2}$ |
| | 1 | $\frac{1}{4}$ | | 3 | $\frac{3}{12}$ | $\frac{1}{4} + \frac{3}{12} = \frac{1}{2}$ |
| | 2 | $\frac{2}{8}$ | | 3 | $\frac{3}{12}$ | $\frac{2}{8} + \frac{3}{12} = \frac{1}{2}$ |
| | 1 | $\frac{1}{3}$ | | 1 | $\frac{1}{6}$ | $\frac{1}{3} + \frac{1}{6} = \frac{1}{2}$ |
| | 2 | $\frac{2}{6}$ | | 2 | $\frac{2}{12}$ | $\frac{2}{6} + \frac{2}{12} = \frac{1}{2}$ |
| | 3 | $\frac{3}{9}$ | | 1 | $\frac{1}{6}$ | $\frac{3}{9} + \frac{1}{6} = \frac{1}{2}$ |
| | 3 | $\frac{3}{9}$ | | 2 | $\frac{2}{12}$ | $\frac{3}{9} + \frac{2}{12} = \frac{1}{2}$ |
| | 4 | $\frac{4}{12}$ | | 1 | $\frac{1}{6}$ | $\frac{4}{12} + \frac{1}{6} = \frac{1}{2}$ |
| | 1 | $\frac{1}{5}$ | | 3 | $\frac{3}{10}$ | $\frac{1}{5} + \frac{3}{10} = \frac{1}{2}$ |

\* Note that the color of the fraction circle pieces will vary depending on the set you use.

## Page 27:

Ten examples are included in the chart below. Accept any other correct student answers.

| First Color * | Number of Pieces | Fraction | Second Color * | Number of Pieces | Fraction | Addition Sentence |
|---|---|---|---|---|---|---|
| | 1 | $\frac{1}{2}$ | | 1 | $\frac{1}{4}$ | $\frac{1}{2} + \frac{1}{4} = \frac{3}{4}$ |
| | 2 | $\frac{2}{4}$ | | 2 | $\frac{2}{8}$ | $\frac{2}{4} + \frac{2}{8} = \frac{3}{4}$ |
| | 3 | $\frac{3}{6}$ | | 3 | $\frac{3}{12}$ | $\frac{3}{6} + \frac{3}{12} = \frac{3}{4}$ |
| | 4 | $\frac{4}{8}$ | | 1 | $\frac{1}{4}$ | $\frac{4}{8} + \frac{1}{4} = \frac{3}{4}$ |
| | 5 | $\frac{5}{10}$ | | 1 | $\frac{1}{4}$ | $\frac{5}{10} + \frac{1}{4} = \frac{3}{4}$ |
| | 6 | $\frac{6}{12}$ | | 1 | $\frac{1}{4}$ | $\frac{6}{12} + \frac{1}{4} = \frac{3}{4}$ |
| | 3 | $\frac{3}{6}$ | | 1 | $\frac{1}{4}$ | $\frac{3}{6} + \frac{1}{4} = \frac{3}{4}$ |
| | 5 | $\frac{5}{10}$ | | 3 | $\frac{3}{12}$ | $\frac{5}{10} + \frac{3}{12} = \frac{3}{4}$ |
| | 1 | $\frac{1}{3}$ | | 5 | $\frac{5}{12}$ | $\frac{1}{3} + \frac{5}{12} = \frac{3}{4}$ |
| | 3 | $\frac{3}{9}$ | | 5 | $\frac{5}{12}$ | $\frac{3}{9} + \frac{5}{12} = \frac{3}{4}$ |

* Note that the color of the fraction circle pieces will vary depending on the set you use.

***Page 28:***

Ten examples are included in the chart below. Accept any other correct student answers.

| First Color * | Number of Pieces | Fraction | Second Color * | Number of Pieces | Fraction | Addition Sentence |
|---|---|---|---|---|---|---|
| | 1 | $\frac{1}{2}$ | | 1 | $\frac{1}{3}$ | $\frac{1}{2} + \frac{1}{3} = \frac{5}{6}$ |
| | 2 | $\frac{2}{4}$ | | 1 | $\frac{1}{3}$ | $\frac{2}{4} + \frac{1}{3} = \frac{5}{6}$ |
| | 3 | $\frac{3}{6}$ | | 1 | $\frac{1}{3}$ | $\frac{3}{6} + \frac{1}{3} = \frac{5}{6}$ |
| | 4 | $\frac{4}{8}$ | | 1 | $\frac{1}{3}$ | $\frac{4}{8} + \frac{1}{3} = \frac{5}{6}$ |
| | 5 | $\frac{5}{10}$ | | 1 | $\frac{1}{3}$ | $\frac{5}{10} + \frac{1}{3} = \frac{5}{6}$ |
| | 6 | $\frac{6}{12}$ | | 1 | $\frac{1}{3}$ | $\frac{6}{12} + \frac{1}{3} = \frac{5}{6}$ |
| | 1 | $\frac{1}{2}$ | | 2 | $\frac{2}{6}$ | $\frac{1}{2} + \frac{2}{6} = \frac{5}{6}$ |
| | 1 | $\frac{1}{2}$ | | 3 | $\frac{3}{9}$ | $\frac{1}{2} + \frac{3}{9} = \frac{5}{6}$ |
| | 1 | $\frac{1}{2}$ | | 4 | $\frac{4}{12}$ | $\frac{1}{2} + \frac{4}{12} = \frac{5}{6}$ |
| | 5 | $\frac{5}{10}$ | | 4 | $\frac{4}{12}$ | $\frac{5}{10} + \frac{4}{12} = \frac{5}{6}$ |

* Note that the color of the fraction circle pieces will vary depending on the set you use.

---

***Page 29:***

One correct answer is given below for each problem. When applicable, accept other student answers (equivalent fractions).

**1.** $\frac{1}{6}$

**2.** $\frac{7}{12}$

**3.** $\frac{11}{10}$

**4.** $\frac{10}{9}$

**5.** $\frac{4}{10}$

**6.** $\frac{2}{6}$

**7.** $\frac{7}{9}$

**8.** $\frac{4}{8}$

**9.** $\frac{14}{12}$

**10.** $\frac{1}{8}$

***Page 30:***

**1.** $\frac{1}{4}$ or $\frac{2}{8}$ or $\frac{3}{12}$

**2.** $\frac{3}{4} = \frac{1}{2} + \frac{1}{4}$ or $\frac{3}{4} = \frac{1}{2} + \frac{2}{8}$

or $\frac{3}{4} = \frac{1}{2} + \frac{3}{12}$

**4.** $\frac{2}{5}$ or $\frac{4}{10}$

**5.** $\frac{1}{2} + \frac{2}{5} = \frac{9}{10}$ or $\frac{1}{2} + \frac{4}{10} = \frac{9}{10}$

**Page 31:**

1. $\frac{1}{12}$

2. $\frac{5}{6}$

3. $\frac{1}{3}$

4. $\frac{1}{2}$ (or $\frac{3}{6}$ or $\frac{6}{12}$ )

   (accept other equivalent fractions)

5. $\frac{1}{2} + \frac{1}{3} = \frac{5}{6}$ (accept other correct student answers)

**Page 32:**

1. $\frac{1}{6}$ (accept other correct student answers)

2. $\frac{1}{8}$

**Page 33:**

When applicable, accept other correct student answers for the problems below.

1. $\frac{1}{4}$

2. $\frac{3}{8}$

3. $\frac{2}{6}$ or $\frac{1}{3}$

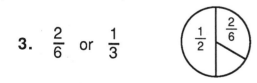

4. $\frac{4}{9}$

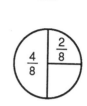

5. $\frac{5}{8} + \frac{3}{8} = 1$

6. $\frac{4}{8} + \frac{2}{8} = \frac{3}{4}$

7. $\begin{array}{r} \frac{3}{12} \\ + \frac{9}{12} \\ \hline 1 \end{array}$

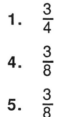

**Page 34:**

1. $\frac{5}{6}$

5. $\frac{3}{6}$ or $\frac{1}{2}$ or $\frac{6}{12}$

6. $\frac{3}{6}$ or $\frac{1}{2}$ or $\frac{6}{12}$

**Page 35:**

1. $\frac{3}{4}$

4. $\frac{3}{8}$

5. $\frac{3}{8}$

**Page 36:**

1. $\frac{7}{12}$

4. $\frac{4}{12}$ or $\frac{1}{3}$

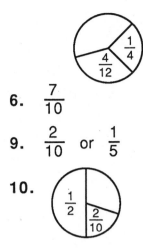

**6.** $\frac{7}{10}$

**9.** $\frac{2}{10}$  or  $\frac{1}{5}$

**10.**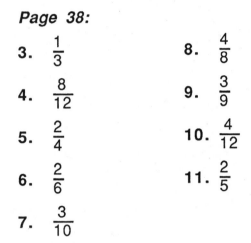

### Page 37:

When applicable, accept other correct student answers to the problems below.

**1.** $\frac{1}{12}$

**2.** $\frac{3}{12}$  or  $\frac{1}{4}$

**3.** $\frac{5}{10}$  or  $\frac{1}{2}$

**4.** $\frac{1}{2} - \frac{1}{8} = \frac{3}{8}$

or $1 - \frac{5}{8} = \frac{3}{8}$

**5.** $\frac{2}{9}$

**6.** $\frac{1}{12}$

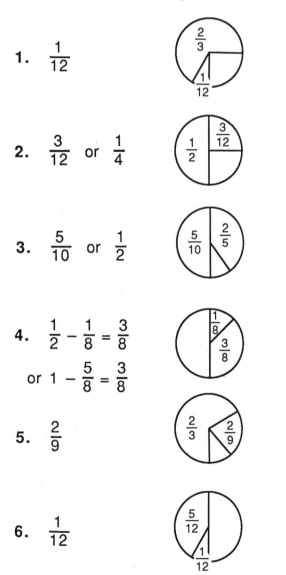

**7.** Accept all correct student answers. Be sure to check that the diagram matches the problem.

### Page 38:

**3.** $\frac{1}{3}$

**4.** $\frac{8}{12}$

**5.** $\frac{2}{4}$

**6.** $\frac{2}{6}$

**7.** $\frac{3}{10}$

**8.** $\frac{4}{8}$

**9.** $\frac{3}{9}$

**10.** $\frac{4}{12}$

**11.** $\frac{2}{5}$

### Page 39:

**1.** $\frac{2}{5}$

**2.** $\frac{3}{12}$

**3.** $\frac{3}{9}$

**4.** $\frac{1}{10}$

### Page 40:

**1.** Orange. (This is the color of the $\frac{1}{12}$ piece on the back cover foldout of this book.)

**2.** twelfth

**4.** 3; $\frac{1}{4} = \frac{3}{12}$

**5.** 2; $\frac{1}{6} = \frac{2}{12}$

**6.**
$$\frac{1}{4} = \frac{3}{12}$$
$$-\frac{1}{6} = \frac{2}{12}$$
$$\rule{3cm}{0.4pt}$$
$$\frac{1}{12}$$

**Page 41:**

1. Green. (Note that the color of the $\frac{1}{6}$ piece will vary depending on the fraction circle set you use.)

2. 6 or sixths

3. $\frac{1}{2} = \frac{3}{6}$;  $\frac{1}{3} = \frac{2}{6}$

4. $\frac{1}{3} + \frac{1}{2} = \frac{5}{6}$

   $\frac{2}{6} + \frac{3}{6} = \frac{5}{6}$

5. $\frac{1}{2} - \frac{1}{3} = \frac{1}{6}$

   $\frac{3}{6} - \frac{2}{6} = \frac{1}{6}$

**Page 42:**

1. Orange. (This is the color of the $\frac{1}{12}$ piece on the back cover foldout of this book.

2. 12 or twelfths

3. $\frac{2}{3} = \frac{8}{12}$;  $\frac{3}{4} = \frac{9}{12}$

4. $\frac{2}{3} = \frac{8}{12}$

   $+ \frac{3}{4} = \frac{9}{12}$

   $\frac{17}{12}$

5. $\frac{3}{4} = \frac{9}{12}$

   $- \frac{2}{3} = \frac{8}{12}$

   $\frac{1}{12}$

**Page 43:**

1. $\frac{2}{3} - \frac{1}{6} = \frac{3}{6}$

   $\frac{4}{6} - \frac{1}{6} = \frac{3}{6}$

2. $\frac{1}{3} = \frac{3}{9}$

   $+ \frac{2}{9} = \frac{2}{9}$

   $\frac{5}{9}$

3. $\frac{4}{6} + \frac{2}{12} = \frac{10}{12}$

   $\frac{8}{12} + \frac{2}{12} = \frac{10}{12}$

4. $\frac{7}{10} = \frac{7}{10}$

   $- \frac{1}{2} = \frac{5}{10}$

   $\frac{2}{10}$

5. $\frac{11}{12} - \frac{3}{4} = \frac{2}{12}$

   $\frac{11}{12} - \frac{9}{12} = \frac{2}{12}$

6. $\frac{6}{8} + \frac{3}{4} = \frac{12}{8}$

   $\frac{6}{8} + \frac{6}{8} = \frac{12}{8}$

**Page 44:**

1. $\dfrac{1}{2} = \dfrac{5}{10}$

   $+\dfrac{2}{5} = \dfrac{4}{10}$

   $\qquad\;\; \dfrac{9}{10}$

2. $\dfrac{5}{10} = \dfrac{1}{2}$

   $+\dfrac{3}{6} = \dfrac{1}{2}$

   $\qquad\;\; \dfrac{2}{2}$ (or 1)

3. $\dfrac{4}{8} = \dfrac{2}{4}$

   $+\dfrac{3}{12} = \dfrac{1}{4}$

   $\qquad\;\; \dfrac{3}{4}$

4. $\dfrac{2}{3} = \dfrac{6}{9}$

   $+\dfrac{5}{9} = \dfrac{5}{9}$

   $\qquad\;\; \dfrac{11}{9}$

5. $\dfrac{3}{5} - \dfrac{1}{2} = \dfrac{1}{10}$

   $\dfrac{6}{10} - \dfrac{5}{10} = \dfrac{1}{10}$

6. $\dfrac{1}{3} + \dfrac{8}{12} = \dfrac{12}{12}$ (or 1)

   $\dfrac{4}{12} + \dfrac{8}{12} = \dfrac{12}{12}$ (or 1)

**Page 45:**

1. $\dfrac{2}{6} = \dfrac{1}{3}$

   $+\dfrac{6}{9} = \dfrac{2}{3}$

   $\qquad\;\; \dfrac{3}{3}$ (or 1)

2. $\dfrac{8}{12} - \dfrac{3}{9} = \dfrac{1}{3}$

   $\dfrac{2}{3} - \dfrac{1}{3} = \dfrac{1}{3}$

3. $1\dfrac{1}{4} = \dfrac{15}{12}$

   $+\dfrac{2}{3} = \dfrac{8}{12}$

   $\qquad\;\; \dfrac{23}{12}$ (or $1\dfrac{11}{12}$)

4. $\dfrac{3}{4} = \dfrac{9}{12}$

   $-\dfrac{2}{6} = \dfrac{4}{12}$

   $\qquad\;\; \dfrac{5}{12}$

5. $\dfrac{3}{6} + \dfrac{4}{8} = \dfrac{2}{2}$ (or 1)

   $\dfrac{1}{2} + \dfrac{1}{2} = \dfrac{2}{2}$ (or 1)

6. $\dfrac{9}{6} = \dfrac{9}{6}$

   $+\dfrac{1}{2} = \dfrac{3}{6}$

   $\qquad\;\; \dfrac{12}{6}$ (or 2)